Bioinformatics in Agriculture and Allied Sectors

Methods and Applications

NIPA® GENX ELECTRONIC RESOURCES & SOLUTIONS P. LTD.
New Delhi-110 034

About the Editors

Dr. P.N. Jagadev, presently working as Professor, Department of Plant Breeding and Genetics, Odisha University of Agriculture and Technology, Bhubaneswar has obtained his M.Sc. (Ag) degree from OUAT, Bhubaneswar, Ph.D. degree from GBPUAT, Pantnagar and Post-Doc. Training from DDPSC, Washington University, St. Louis, USA. He also served as the Dean of Research, Dean of Extension Education, OUAT, Bhubneshwar, Orissa, Professor and Head, Department of Bioinformatics of Odisha Agrivarsity. He has research areas of specialization in plant stress breeding, biotechnology and bioinformatics with more than 32 years of distinguished career on teaching and research. He travelled many countries like USA, Australia, South Korea, Hong Kong, Thailand and Kenya on different R&D assignments. He has guided 55 students of PG, Ph.D. and PDF levels. To his credit, he has 254 publications (research papers-83, conference/ symposium papers–40, books/manuals-32, book chapters-9, technical booklets-25 and popular scientific articles-65). He is a Fellow of Indian Society of Genetics and Plant Breeding since 1986. He has also developed five crop varieties of rice, sugarcane, jute and mesta for the farmers of Odisha. He has the distinction of receiving several awards like Krutti Krushi Vaigyanik Samman and Best Agricultural Scientist Award from the Government of Odisha, Krushi Ratna Award from Krushak Sambad and Best Director Award from Education Expo TV, Mumbai.

Dr. Sashikala Beura, presently working as the Professor and Head, Department of Floriculture and Landscaping and as Director, Biotechnology-cum-Tissue Culture Centre, Odisha University of Agriculture and Technology, Bhubaneswar, has obtained her M.Sc. (Ag) degree in Horticulture from OUAT, Bhubaneswar with University Gold Medal and Ph.D. degree from GBPUAT, Pantnagar with CSIR Sr. Research Fellowship. She has research areas of specialization in Floriculture, Landscaping, Biotechnology and Bioinformatics with more than 27 years of distinguished career on teaching and research. She travelled many countries like Australia, Thailand and Iran on different R&D assignments. She has guided 33 PG and 2 Ph.D. Students. To her credit, she has 255 publications (research papers-79, conference/symposium papers–78, books/manuals-8, book chapters-9, technical booklets-15 and popular scientific articles-66). She has the distinction of receiving several awards like Alaka Manjari Rath Best Woman Scientist Award from the Governor of Odisha, Best Faculty Award from Education Expo TV, Kolkata and two best paper Awards. She has also developed a cashewnut hybrid 'Jagannath' and introduced the protected cultivation of gerbera, lilium and anthurium for the farmers of Odisha.

Bioinformatics in Agriculture and Allied Sectors

Methods and Applications

P.N. Jagadev
Professor & Ex-Head (Bioinformatics)
Department of Plant Breeding and Genetics
Odisha University of Agriculture and Technology
Bhubaneswar, Odisha

Sashikala Beura
Professor & Head
Department of Floriculture & Landscaping and
Director, Biotechnology-cum-Tissue Culture Centre
Odisha University of Agriculture and Technology
Bhubaneswar, Odisha

NIPA® GENX ELECTRONIC RESOURCES & SOLUTIONS P. LTD.
New Delhi-110 034

NIPA® GENX ELECTRONIC
RESOURCES & SOLUTIONS P. LTD.

101,103, Vikas Surya Plaza, CU Block
L.S.C. Market, Pitam Pura, New Delhi-110 034
Ph : +91 11 27341616, 27341717, 27341718
E-mail: newindiapublishingagency@gmail.com
www: www.nipabooks.com

For customer assistance, please contact
Phone: + 91-11-27 34 17 17
Fax: + 91-11- 27 34 16 16
E-Mail: feedbacks@nipabooks.com

ISBN: 978-81-19002-90-0

Composed and Designed by NIPA®.

Dedicated to

SHRESTHAA

Our Loving Daughter

त्रिलोचन महापात्र, पी.एच.डी
एफ.एन.ए.,एफ.एन.ए.एस.सी., एफ.एन.ए.ए.एस
सचिव एवं महानिदेशक

TRILOCHAN MOHAPATRA, Ph.D
FNA, FNASc, FNAAS
SECRETARY & DIRECTOR GENERAL

भारत सरकार
कृषि अनुसंधान और शिक्षा विभाग एवं
भारतीय कृषि अनुसंधान परिषद
कृषि एवं किसान कल्याण मंत्रालय, कृषि भवन, नई दिल्ली 110001

GOVERNMENT OF INDIA
DEPARTMENT OF AGRICULTURAL RESEARCH & EDUCATION
AND
INDIAN COUNCIL OF AGRICULTURAL RESEARCH
MINISTRY OF AGRICULTURE AND FARMERS WELFARE
KRISHI BHAVAN, NEW DELHI 110 001
TEL: 23382629, 23386711 FAX: 91-11-23384773
E-mail: dg.icar@nic.in

Foreword

Bioinformatics is a recent and emerging interdisciplinary field of science combining biology, computer science and information technology to develop methods and software's for analyzing large data sets. Several mathematical and statistical tools are used through computational applications and development of algorithms to extract maximum knowledge from biological data. Bioinformatics is gaining attention in academic and industrial sectors due to its utilization in land use planning, impact assessment and molecular biology application in plant breeding. It has a tremendous scope in scientific research and entrepreneurship as its use is rapidly increasing in India and abroad.

In this context, this book on "Bioinformatics in Agriculture and Allied Sectors" is a timely contribution of editors with an excellent collection of several aspects of bioinformatics and biotechnology. The book contains important areas of soil informatics, proteomics, nanotechnology, biosensors, sequence alignment, homology modeling, codon usage, drug designing and application of bioinformatics tools in aquaculture, avian sectors and design of earthquake resistant agricultural structures. The editors focused on the application of computation and software's for the effective use of databases through bioinformatics methods and tools. The contents of this book are informative and would be highly useful for the students and scientists of biological sciences including clinical personnel.

I appreciate the efforts of entire team and congratulate Dr. P.N. Jagadev, Professor (Plant Breeding and Genetics) and Ex-Head (Bioinformatics) and Dr. Sashikala Beura, Professor and Head (Floriculture and Landscaping) and Director, Biotechnology-cum-Tissue Culture Centre of Odisha University of Agriculture and Technology, Bhubaneswar for their creditable task in editing this book, which comprehensively summarizes the latest scientific findings and methods in bioinformatics. I am sure that the contents provided in the various chapters of this book by authors will be a valuable resource and useful to all the professionals, scientists and students concerned with the bioinformatics education and research in agriculture and allied sectors.

Dated the 6th September, 2018
New Delhi

T. Mohapatra

Preface

Bioinformatics, a recently emerged frontier discipline, is the convergence of analytical and computational tools with the disciplines of biological sciences. Now-a-days, it has a great scope for development of entrepreneurship. The integration of three biological processes namely, DNA sequence determining protein sequence, protein sequence determining protein structure and protein structure determining protein function, makes us to achieve the long term goal of the complete understanding of the biology of organisms through the development of bioinformatics tools and software. Government of India is also giving much thrust for the development of this area through the Department of Biotechnology.

The present book "Bioinformatics in Agriculture and Allied Sectors" contains 26 chapters including some fundamental chapters of soil informatics, proteomics, sequence alignment, homology modeling, drug designing and bioinformatics application in aquaculture and avian sectors. Efforts were given to make this book in a simplified way for well understanding and to make very useful for the students, scientists and professionals engaged in research and development of bioinformatics education.

We feel immense pleasure to express our heartfelt gratitude to Dr. T. Madhan Mohan, Adviser, Department of Biotechnology, Ministry of Science and Technology, Government of India for his inspiring guidance, encouragement and support for organizing three national workshops-cum-training on bioinformatics at OUAT and this book is an outcome of that.

We are highly grateful to Dr. T. Mohapatra, Secretary, DARE and Director General, ICAR, Government of India for writing a foreword for this book.

We also like to acknowledge the support and encouragement received from Dr. R.K. Mahapatra, Chief Librarian, OUAT and appreciate the efforts of M/s New India Publishing Agency, New Delhi for publishing this book in a beautiful form.

We wish to express our sincere gratitude to the authors, who contribute the manuscripts to make this publication, a success.

Last but not the least, we express our appreciation for our loving daughter Shresthaa for her understanding and patience during the long hours we spent in the preparation of this book.

Dated the 11th September, 2018
Bhubaneswar

P.N. Jagadev
Sashikala Beura

Contents

List of Contributors

Barik, B.P.
Post-Graduate Department of Bioinformatics, North Orissa University, Baripada
Mayurbhanj-757003

Beura, Sashikala
Biotechnology-cum-Tissue Culture Centre, Odisha University of Agriculture and Technology
Bhubaneswar-751003

Dash, B.P.
School of Biotechnology, Fakir Mohan University, Balasore-756019

Dash, R.
Department of Bioinformatics, Odisha University of Agriculture and Technology
Bhubaneswar-751003

Dhal, N.K.
Department of Plant Pathology, Odisha University of Agriculture and Technology
Bhubaneswar-751003

Gouda, L.
Department of Water Resources, Government of Odisha, Bhubaneswar-751001

Jagadev, P.N.
Department of Plant Breeding and Genetics, Odisha University of Agriculture and
Technology, Bhubaneswar-751003

Kanungo, L.
Department of Bioinformatics, Odisha University of Agriculture and Technology
Bhubaneswar-751003

Martha, S.R.
Department of Bioinformatics, Odisha University of Agriculture and Technology
Bhubaneswar-751003

Mishra, B.K.
Department of Soil Science and Agricultural Chemistry, Odisha University of Agriculture and
Technology, Bhubaneswar-751003

Mishra, L.D.
Department of Plant Breeding and Genetics, Odisha University of Agriculture and Technology
Bhubaneswar-751003

Misra, A.N.
School of Biotechnology, Fakir Mohan University, Balasore-756019

Misra, M.
Institute of Frontier Science and Biotechnology, Baramunda, BDA Colony, Bhubaneswar-751003

Mohanty, S.
Department of Bioinformatics,Odisha University of Agriculture and Technology Bhubaneswar-751003

Mohapatra, A.P.K.
Department of Animal Physiology, Odisha University of Agriculture and Technology Bhubaneswar-751003

Nayak, R.
National Institute of Science Education and Research, Bhubaneswar

Padhy, S.
School of Mathematics, Statistics and Computer Science, Utkal University Bhubaneswar-751004

Panda, P.K.
Natural Products Department, Institute of Minerals and Materials Technology Bhubaneswar-751013

Parija, S.C.
Department of Pharmacology and Toxicology, Odisha University of Agriculture and Technology Bhubaneswar-751003

Patra, R.R.
Department of Master in Computer Application, Odisha University of Agriculture and Technology, Bhubaneswar-751003

Pradhan, B.
Department of Plant Breeding and Genetics, Odisha University of Agriculture and Technology Bhubaneswar-751003

Pradhan, S.K.
Department of Bioinformatics, Odisha University of Agriculture and Technology Bhubaneswar-751003

Rath, S.N.
Department of Bioinformatics, Odisha University of Agriculture and Technology Bhubaneswar-751003

Ray, D.P.
Ex-Vice-Chancellor, Odisha University of Agriculture and Technology, Bhubaneswar-751003

Roy, A.K.
Bioinformatics Centre, ICAR-Central Institute of Freshwater Aquaculture, Kausalyaganga Bhubaneswar-751002

Sarangi, N.
Ex-Director, ICAR-Central Institute of Freshwater Aquaculture, Kausalyaganga Bhubaneswar-751002

Sardar, K.K.
Department of Pharmacology and Toxicology, Odisha University of Agriculture and Technology Bhubaneswar-751003

1

Bioinformatics and Its Relevance

P.N. Jagadev

Bioinformatics is a recently emerged frontier discipline, which helps in analyzing biological data with the help of information technology and statistical software packages. In a broader sense, bioinformatics is the application of computer technology in management of biological information. This interdisciplinary scientific tool having no barriers between different branches of science like biology, biotechnology, mathematics, computer science and information technology, has great influence on advanced bio-science research, particularly dealing with genes, proteins and cell biology.

Bioinformatics is the branch of study that deals with the use of information technology (IT) for managing data of biological science. Bioinformatics drew public attention with the announcement of the draft sequence of human genome by the then US President Bill Clinton along with Craig Venter and Francis Collins in 2000 (Uma and Sathyanathan, 2004).

Bioinformatics and life sciences research is gaining attention in the Government, industries and academic sectors. India was the first country in the world to establish a biotechnology information system (BTIS) network in 1987 through the Department of Biotechnology and created an infrastructure that enables it to harness biotechnology through the application of bioinformatics (Sibal, 2006). Through this branch of science, biological data are acquired, stored and then analyzed. Not only cloning, it has become a specialized branch in areas such as pharmaceuticals, agrochemicals, hydrocarbon compounds and derivatives and environmental research. It deals mainly with the development of IT tools that help to achieve new visions in areas surrounding molecular biology. By this, one can search answers to many genetic diseases like sickle cell anemia, why one drug works better on a particular person than others and how to grow healthier varieties of plants. Hence, person having combined skills of biology and IT, have ample opportunities in this area. Now, almost all major global IT giants have set up laboratories in India to develop a protein database and undertake research.

The development of molecular biology reminds us that during the first quarter of 20th century, it was found out that cellular basis of information was chromosomes, in the 2nd quarter, biologists discovered the molecular basis of information was DNA, in the 3rd quarter, recombinant DNA tools were developed and in the 4th quarter of the century, genetic information from genes and their entire genomes were gathered. The enormous biological data generated in recent years led to the necessity of storage of such data for future use; need for radical new methods for analyzing these databases and need for powerful hardware to carry out the task of analyzing these databases.

Thus, the relevance of bioinformatics came in. There were developments in protein and gene sequencing in 1970s and the use of computers gained popularity for data storage and access to information from mid-1980s and onwards. With the introduction of worldwide web (www) in 1990s, the developmental work in bioinformatics was given a boost (Senapati, 2005).

Applications

Bioinformatics or the science of informatics as applied to biological research, has grown rapidly. Bioinformatics can be used for applications in various fields like molecular medicine, preventive medicine, gene therapy, drug development, antibiotic resistance, microbial genome studies, improvement of nutritional quality, waste cleanup, alternative energy sources, crop improvement, development of biotic and abiotic resistant varieties, climate change studies, comparative studies, evolutionary studies, biotechnology and veterinary science etc. (Seth, 2005). Bioinformatics is a multidisciplinary approach that combines computational and biological expertise to analyze biological data (both genomic and clinical) to advance research and development in life sciences (Ghose, 2006).

We need bioinformatics to organize and integrate the information for analysis and interpretation of various types of data including nucleotide and amino acid sequences, protein domains and structures. This helps us to know about the genes connected to a disease, proteins encoded by them, conditions under which they expressed, pathways in which they participate and the targets for new therapeutics. Bioinformatics symbolizes the convergence of biology and IT for prosperity. It not only offers applications in areas like molecular medicine, microbial genomes, risk assessment, DNA forensics, but also in agriculture, livestock and fishery breeding and bio processing. Understanding plant and animal genomes will allow us to create stronger, more disease resistant plants and animals, reducing the costs of agriculture and providing consumers with more nutrition, pesticide-free foods through the use of bio pesticides, producing edible vaccines incorporated into food products.

Bioinformatics can be used to store huge bioscience data for their future uses. By exploiting computational tools, databases and new strategies for pests and diseases, resistant genes can be identified and plants with desired traits can be engineered to enhance agricultural productivity. Bioinformatics can be utilized in two ways for agriculture, viz. (i) at molecular level for application in breeding strategies and (ii) agri-informatics for land use planning, farmers' window, early warning system, impact assessment and environmental applications.

Opportunities

The bioinformatics industry in India is growing very rapidly as is the case with the other countries in the region. The industry has developed several highly useful new bioinformatics tools and databases in the recent past (Kolaskar, 2006). The global market for bioinformatics products and services is estimated to be around US $ 5 billion. The major activities in global arena include major areas like education and training, R & D, providing research facilities and commercialization. The limited domestic market size and extent of profitability has made many domestic players to look at global opportunities. The Indian bioinformatics industry is estimated at Rs. 350 crores growing at rate of 35% per annum and will provide tremendous opportunities for professionals and companies (Sinha, 2005).

Job prospects are in all sectors of biotechnology, pharmaceutical, biomedical sciences, agriculture, research institutions, hospitals and information technology industries. Industries where trained bioinformatics specialists are in demand are pharma companies, IT companies, chemical industries, R & D organizations, agricultural sectors, plastic industries, food industries, petroleum derivatives and academic institutions etc. (Philip, 2006).

Outlook

Increased interaction among bioscientists and information scientists is needed to allow understanding of each others strengths and requirements. Agricultural scientists in biosciences should come forward and acquaint their counterparts in IT with their problems in information requirements and processing of biological datasets. Hence, although there is wide application of IT in bioinformatics, the input data have to come from biology and biologists.

The Indian entrepreneurs need to develop its knowledge base as Indians can not afford to outsource its research requirements. They also need to build a team of bioinformaticians that can perform computational analysis and also thoroughly seasoned in the intricacies of biological sciences. Many of biopharma players have in-house expertise and even in case of those willing to outsource, there needs to be considerable efforts to convince such. The future of bioinformatics, though impossible to predict, seems very bright. This is an amalgamation of so many highly technical fields that any new development in one will give bioinformatics a direct push. At one end, life science research is getting more and more sophisticated and at the other end, the computers are getting better everyday and the technologies related to networking are constantly evolving. This will have a tremendous impact on bioinformatics, which will have a multifactor growth leading it to evolve as the mainstream science of this century.

Acknowledgement

The author duly acknowledges the Central Institute of Freshwater Aquaculture, Bhubaneswar for the support to prepare this manuscript.

References

Ghose, J. 2006. Fact file on competitive examinations-bioinformatics entrance examination. *Employment News*. (9-15 Sept., 2006). 31 (24): 1.

Kolaskar, A.S. 2006. Foreword. *Proc. Internatl. Conf. Bioinformatics (InCoB, 2006)*. Dec. 18-20. New Delhi.

Philip, S.A. 2006. Science in farm. Education Express. *The New Indian Express*. Nov.17. pp.1.

Senapati, B. 2005. Bioinformatics- a new face of science. *Proc. Natl. Seminar on Frontier Bioinformatics Tools in Genome and Proteome Analysis,* (March 14-15), OUAT, Bhubaneswar. pp.56.

Seth, P.K. 2005. Bioinformatics: global perspectives. In: *Science and Technology for Future* (Ed. Seema Shah). Neeta Press Publ., New Delhi, pp. 69-80.

Sibal, K. 2006. Message. *Proc. Internatl. Conf. Bioinformatics (InCoB, 2006)*. Dec. 18-20. New Delhi.

Sinha, R. 2005. When bio meets IT. *The Times of India. Bhubaneswar-National.* Aug. 2. 2005. pp.14.

Uma, G. and Sathyanathan, B. 2004. What is bioinformatics? *Sci. Reporter*. Nov. issue. pp. 39-42.

2

Bioinformatics in Agriculture and Allied Sectors

D.P. Ray

Bioinformatics is a multidisciplinary approach that combines computational and biological expertise to analyze biological data to advance research and development and it drew public attention with the announcement of the draft sequence of human genome by the then US President, Bill Clinton along with Craig Venter and Francis Collins in 2000. Bioinformatics and life sciences research are gaining attention in the government, industries and academic sectors. India was the first country in the world to establish a biotechnology information system (BTIS) network in 1987 through the Dept. of Biotechnology and created an infrastructure that enables it to harness biotechnology through the application of bioinformatics. Through this branch of science, biological data are acquired, stored and then analyzed. Bioinformatics can be used to store huge bioscience data for their future uses. By exploring computational tools, databases and new strategies for pests and diseases, resistant genes can be identified and plants with desired traits can be engineered to enhance agricultural productivity. Bioinformatics can be utilized in two ways for agriculture, viz., (i) at molecular level for application in breeding strategies and (ii) agri-informatics for land use planning, farmers window, early warning system, impact assessment and environmental applications.

At present, the Indian bioinformatics industry is doing a business with an investment of Rs. 350 crores. The business is growing at a rate of 35% per annum and will provide tremendous opportunities for professionals and companies in India.

The future of bioinformatics seems very bright. This is an amalgamation of so many highly technical fields that any new development in one will give bioinformatics a direct push. At one end, life science research is getting more and more sophisticated and at the other end, the technology of computers is getting better day by day and the technologies related to networking are constantly evolving. This will have a tremendous impact on bioinformatics, which will have a multifactor growth leading to evolve as the mainstream science of this century.

The integration of information is required for complete understanding of key biological processes operating in living organisms and bioinformatics has a great role to play in this direction.

The scientists have succeeded in transferring genes into rice to increase levels of Vitamin A, iron and other micronutrients. Progress has been made in developing cereal varieties that have a greater tolerance to soil alkalinity, free aluminium and iron toxicities.

Similarly genes resistant to various biotic and abiotic stresses can be identified. Varieties thus developed will address problems of malnutrition of masses and improve crop productivity in problem soils. It is expected to come up with much more innovations with the use of biotechnological and bioinformatics tools.

3

Soil Informatics of Odisha

B.K. Mishra

Soil is formed by disintegration and decomposition of rocks, minerals and organic matter, Soil is a habitant for plant growth and serves as a storehouse of water and food nutrients for plants, Soil is not same everywhere and varies morphologically, physically, chemically and mineralogically from one point to another both in horizontal and vertical directions. Development of soil informatics is necessary to classify the soils, to assess the productivity of soils for various crop productions and to use the soil for various purposes such as land capability classification, soil irrigability classification, soil rating and soil mapping.

Morphologically soil may appear red, yellow, black, brown or ashy in colour. Soil may vary from sandy to clayey in texture. Depending upon the nutrient reserve capacity, the soil may be fertile or infertile. On the basis of climate, a soil may be classified as laterite soil. Red earth and red loam soil, podzol soil or desert soil.

Different kinds of soil contain minerals of various types and amounts and so exhibit distinct physio co-chemical properties. Although various methods have been used in the past to classify the soils, the most modern and scientific methods is based on dominants pedogenic processes acting on a place to from a soil. The new method of soil classification has two important characteristics-Firstly soil classification is based on certain diagnostic parameters which could be estimated quantatively, and secondly depending on diagnostic parameters, a scientific method is being followed to classify a soil from higher to lower levels such as order, suborder, great group, subgroup, family and series.

As per the modern system, all soils of the world have been classified into eleven classes at order level such as Alfisol, Andisols, Aridisol, Entiosol, Histosol, Inceptisol Mollisol, Oxisol, Spodosol, Ultisol and Vertisol. Soils among themselves are classified on the basis of presence or absence of diagnostic surface or subsurface horizons. The diagnostic surface horizons may be Anthropic, Histic Mollick, Mollic, Ochric, Plaggen, Umleric epipendons. Similarity the diagnostic sub-surface horizons are Agric, Albic, Argillic, Calcic, Cambic, Duropan, Fragipan,

Gypsic, Kandic, Oxic, Salic, Spodic and Sulfuric horizons. Other diagnostic parameters that are used to classify a soil at lower levels are classes of soil moisture regimes (Aquic, aridic, udic, ustic and xeric), soil temperature regimes (Pergelic, cryic, frigid, mesic, thermic and hyperthermic).

Development of Soil Informatics: It is necessary to develop a data base system to collect data on soil properties from different places and then process the data to find out the presence and absence of diagnostics parameters. A surface package is to be developed to classify the soil from its higher level to lower levels. As an examples a flow chart is presented below to prepare a software package so as to classify a soil at order level:

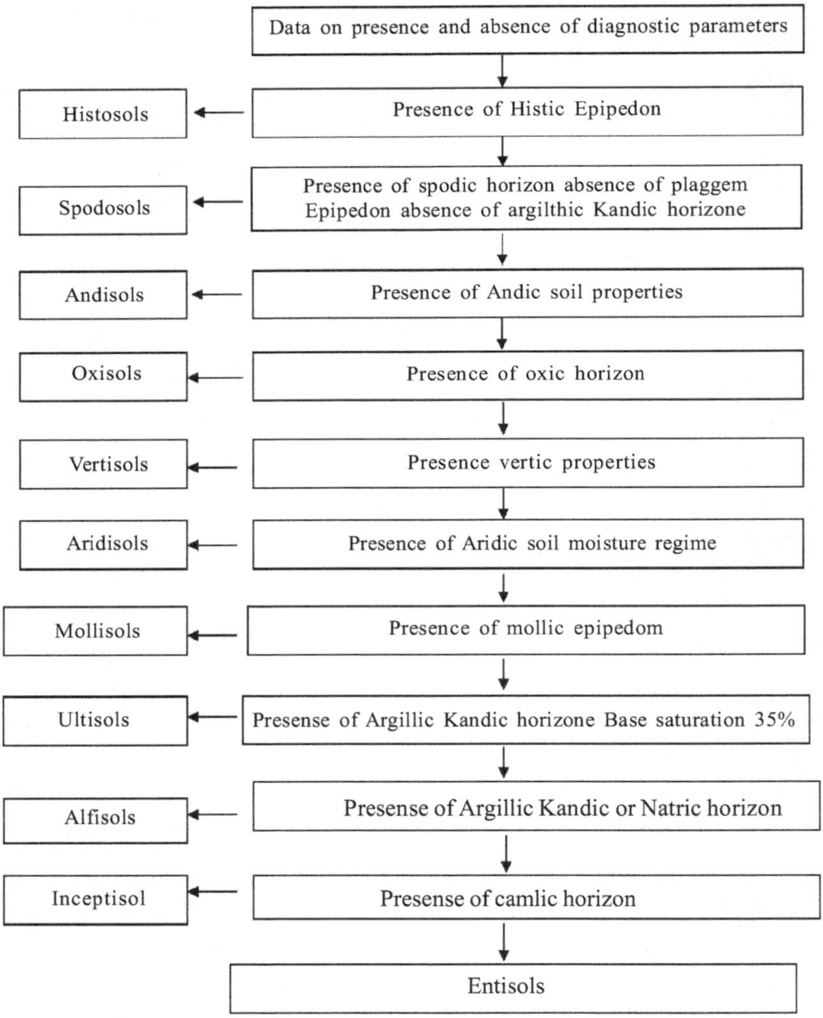

Development of Soil Informatics for Odisha

So far now, four kinds of soils such as Alfisol, Inceptisol, Vertisol and Entisols have been identified at order level .These soils are further divided into nine subclasses at suborder level and into fifteen subclasses at great group level. Classification of these soils at still lower level becomes a voluminous work on the parts of scientists or the users of soil data. As the examples, the developments of a soil informatics for Alfisol a dominants soil group of Odisha, is presented below.

A. Classification of Alfisol at sub-order level

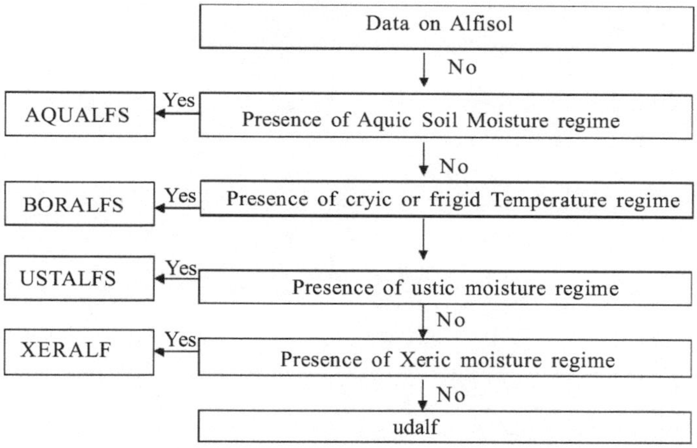

B. Classification of USTALF at great group level

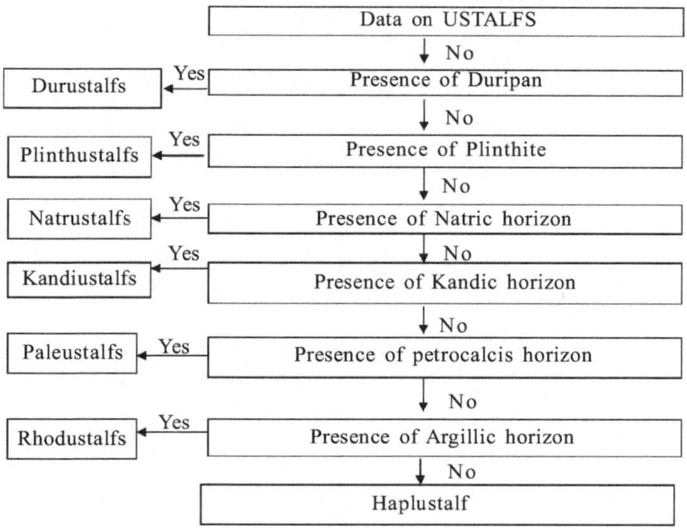

Use of Soil Informatics

Soil informatics finds several uses not only for soil scientists agriculture officers and farmers but also for the planners at state or National level. Soil informatics would be an easy access for a soil scientist to correlate the soils of the world, to identify soils, if present, at some places and to prepare land use planning. It also helps the Agriculture officers and farmers to identify the soil constraints that limit crop production and to access the soil productivity for various crop production. The planners may use the soil informatics to make strategic planning on food, filore or fuel production at regional, state or national level.

4

Use of Bioinformatics Tools and Techniques in Agriculture

S.N. Rath and S. Mohanty

Advancement of both biological sciences and computer technology gave birth today's one of the hottest area known as bioinformatics. Gradually bioinformatics analysis has become an essential component to the continued improvement and development of any kind of biological research work. It has various applications in medicine, biotechnology, agriculture etc. Some of the applications of bioinformatics related to biological information analysis are as follows [1]

- Information related to bio-molecules can be mapped (e.g.), the sequences can be parsed to find sites where so-called "restriction enzymes" will cut them.
- Sequences can be compared, usually by aligning corresponding segments and looking for matching and mismatching letters in their sequences. Genes or proteins that are sufficiently similar are likely to be related and are therefore said to be "homologous" to each other.
- If a homologous exists then a newly discovered protein may be modeled - that is 3D structure of the gene product can be predicted without doing laboratory experiments.
- Bioinformatics is used to attempt to predict the function of actual gene products.
- Information about the similarity, and by implication, the relatedness of proteins is used to trace the "family trees" of different molecules through evolutionary time.
- Molecular modeling /structural biology is growing field, which can be considered part of bioinformatics. There are, for example, tools which allow making predictions of the secondary structure of proteins arising from given amino acid sequence. These are often based on known "solved" structures and other sequenced molecules acquired by structural biologists.

- Structural biologists use bioinformatics to handle the vast and complex data from X-ray crystallography, nuclear magnetic resonance (NMR) and electron microscopy investigations and create the 3-D models of molecules.
- Information of the research records or the research papers of the journal are stored in text search database for further reference to enhance and improve the research procedure or to get the information for related research techniques as a reference.
- There are other fields like medical imaging / image analysis that might be considered as a part of the bioinformatics. There is also a whole other discipline of biologically inspired computation: genetic algorithms, AI, neural networks etc. Often these areas interact in unexpected ways. Neural networks, inspired by crude models of the functioning of the nerve cells in the brain, and used in a program called PHD to predict, the secondary structures of proteins from their primary sequences.
- Agro-informatics or Agricultural informatics: concentrates on the aspects of bioinformatics dealing with plant genomes.

Retrieval of biological information from biological databases is important for the biotechnological and bioinformatics professionals, since it is a scholarly need, in accordance requirement and means of updating oneself professionally. Now a day, information technology, communication networks, Internet and bioinformatics websites offer the biotechnological and bioinformatics professionals, the timely and convenient access of the required information from biological databases. The life science librarians have realized that there is a need to organize the information available on the web so that biological research professionals can easily find information to enrich the education process, facilitate the basic research and improve the research skills. The internet has provided the new ways for the research scholars to interact with machine and communicate with the fellow colleagues in their day to day research skills, research procedure and decision making. The researchers irrespective of geographical locations, languages and cost can access the bioinformatics, biotechnology and medical websites.

A biological database is a large, organized body of persistent data, usually associated with computerized software designed to update, query, and retrieve components of the data stored within the system. A simple database might be a single file containing many records, each of which includes the same set of information.

For example, a record associated with a nucleotide sequence database typically contains information such as contact name; the input sequence with a description of the type of molecule; the scientific name of the source organism from which it was isolated; and often, literature citations associated with the sequence. For researchers to benefit from the data stored in a database, two additional requirements must be met:

- Easy access to the information.
- A method for extracting only that information needed to answer a specific biological question.

A wide range of database is present to get the information in text format. Mostly the databases of Biological Research Literature are text based search. Some of the text based database search includes [1].

- Generalized Databases (DNA, Proteins and Carbohydrates, 3D-Structures).
- Specialized Databases (Genomes, Protein Families, Pathways, Microarray data etc.)
- Various methods are available to search a database.
- Text-based database search (SRS, Entrez etc).
- Sequence- based database search (sequence similarity search, BLAST, FASTA etc.).
- Motif-based database search (ScanProsite, Emotif).
- Structure-based database search (Structure similarity Search, VAST, DALI etc.).

A wide range of database is present to get the information in text format. Mostly the database of Biological Research Literature is text based search. Some of the text based database search includes

- ENTREZ (Integrated database search for nucleotides, proteins, genomes, structures, populations sets and literature at NCBI, USA).
- SRS6 (Search EMBL, SWISS-PROT, TrEMBL, PIR, PDB etc. at EBI, UK, or other SRS5 Servers world wide).
- DBGET/LinkDB (Integrated database search of Link DB at GenomeNet, Kyoto University and University of Tokyo).
- SSS (Sequence and structure searching site at Berkeley, US).

- Molecular R US (Full text search of the PDB database at NIH).
- MIA (Molecular Information Agent searching biological databases to find the existing information about a macromolecule at SDSC, USA).

The internet not only offered global electronic journals, but it provides the latest review of information. The www as the most revolutionary development in publishing arena since the printing press. It is evident that electronic version of journals has some inherent advantages over print version such as incorporation of multimedia, liberal use of colors, elimination of delay in manuscript processing, news alerting, instant real time online discussion. The use of Internet in the research process begins from identifying research issues, through using the web for survey, to pre-publishing and publishing results. The literature searches using different databases such as Medline are obviously an important and integral part of every research process. Material published on the Internet government, biotech organizations, other institutions, medical journals and pharmaceutical companies as well as through medical web directories. The medical professionals are able to retrieve the information from these websites in order to make a better decision of research procedure. Many research institutions are now having research records on the websites for research and education purposes.

Bioinformatics involves promoting cutting-edge scientific research for development of beneficial new tools and products that seek new and better ways to sustain and improve agriculture, to protect environment, and to generate new economic opportunities for agricultural communities [3]. Bioinformatics has offered an invaluable tool to not only understand the epidemiology and evolution of virulent forms of APMV-1, but also to develop and improve diagnostic tests. Department of Agriculture (USDA) validated a real-time reverse transcription-PCR (RRT-PCR) test (Fusion test) directed at the fusion-cleavage site of NDV, the goal was to differentiate virulent Newcastle disease virus strains from those of low virulence. Bioinformatics tools such as comparative nucleotide alignment and phylogenic analysis used to identify how this isolate differed from other detectable isolates, to identify other isolates that may fail detection, and to determine which mismatches allow the virus to escape detection [2].

Modern agricultural biotechnology makes use of genetic engineering techniques, the alteration of genetic material through molecular biology, to modify living organisms such as plants, animals, and microbes. Today, genomics, the study of all the genetic material in an organism, is leading to tremendous advances in biotechnology. Genomics is both generating new tools and techniques and producing huge amounts of biological data. The deluge of genomic data has even led to the new science of bioinformatics, which enables the data to be stored, accessed, compared, and used. As a result of genomics, genes for desirable traits can be rapidly identified and used to create new biotechnology products. Bioinformatics also helps scientists to predict which part of a genome encodes a desirable trait. Using bioinformatics tools, scientists can search genomic data and identify a region important for a desired trait; then, through biotechnology methods, transfer that trait to another organism to create a useful product or outcome, for example converting a drought-sensitive crop to a drought-tolerant crop. Bioinformatics, then, allows genomic data to be used quickly and effectively in biotechnology [4].

References

[1] S.C.Rastogi, Namita Mendiratta, Parag Rastogi, Bioinformatics Concepts, Skills &Applications, CBS Publication, Edition 2004.

[2] http://ars.usda.gov/research/publications/publications.htm SEQ_NO_115=200002

[3] http://www.csrees.usda.gov/nea/biotech/biotech_all.html\parpard\sb100\sa100\f0\fs24 Overview\par

[4] http://www.hyperdictionary.com/search.aspx?Dict=define=biotechnology\cf0\none

5

Proteomics in Crop Improvement for Drought Tolerance

P. N. Jagadev

If the 1990s were the decade of genomics, the first ten years of the new century are set to become the decade of proteomics. For the first time, the technologies of proteomics make it possible to generate quantitative protein expression data on a scale and sensitivity comparable to that achieved at the genetic level. This advance has major implications for our understanding of cellular organisation in health and disease and for pharmaceutical and agricultural biotechnology. Indeed, proteomics is already yielding important findings, across a wide range of applications. Most proteomics research, however, is directed towards the more proximal goal of investigating protein expression and function under specified physiological conditions.

The rapid evolution of proteomics has continued during the past years with a series of innovations in the core technologies of two-dimensional electrophoresis and mass spectrometry, and a diversity of productive research programmes. Well annotated proteomics databases are now emerging in a number of fields to provide a platform for systematic research, with particularly promising progress in clinical applications such as cardiology and oncology. Large-scale quantitative research, comparable in power and sensitivity to that achieved for gene expression, is thus becoming a reality at the protein level.

Proteomics is the study of an organism's entire complement of proteins. It is the separation of all the proteins of a cell, organism or biological fluid including the assessment of the expression level in a sample and the determination of the identity of a protein of interest. Basically, proteins provide structural and functional framework for cellular life. Genetic information is static while the protein complement of a cell is dynamic. Proteomics is one of the fastest-growing areas of biological research now. Australians pioneered the science and the term "proteome" was coined by Marc Wilkins of Macquarie University, Australia in 1994.

Proteomics lies within the realm of the life sciences, which have been recognized as an important area of contribution to economic growth. Technological developments in proteome analysis have the potential to influence the productivity and efficiency of many areas associated with the life sciences, including the agricultural, pharmaceutical and medical industries. Proteins of interest to the agricultural industry include those responsible for crop stress-tolerance, those which confer specific traits related to increased crop yields and proteins which provide insight into animal diseases.

Proteomics as a Tool for Crop Improvement

Plant biologists have used this technique for various purposes, including for genetic or physiological studies. The opportunity to reveal hundreds of gene products on one single gel has permitted examination of the differences in protein patterns between genotypes, and in this way genetic distances have been estimated and phylogenetic relationships established. Variations in gene expression have also been studied during development and in response to various treatments and this has led to the identification of the proteins whose expression is regulated during these conditions. Mutants have been characterized at the protein level and pleiotrophy estimated. The genes encoding the identified proteins have been mapped and their products show variation in isoelectric point or molecular mass (i.e. position shift or PS variants). The expressed genes can be added to the genetic maps currently established with nucleic acid markers.

The recently developed 'transcriptome' tools, such as high density cDNA filters, cDNA filters, cDNA microarrays or DNA chips will be usefully complemented by proteomics, although it is noteworthy that the amounts of a protein and of its mRNA are not always correlated Further, since protein turnover and post translational modifications can not be studied at the cDNA or mRNA level, the importance of on-going proteomics is and will continue to be critical importance.

Water deficit induces the expression of proteins not specifically related to this stress, but rather to reactions against cell damage. These include different classes of heat shock protein genes, thiol proteases, proteinase inhibitors and osmotin. In maize, a ferritin gene induced by iron stress is also induced by drought and ABA. Several genes encoding proteins with functions not directly related to stress were shown to be expressed at greater levels in response to drought or salinity stress. It has been reported that cell wall hardening, a phenomenon occurring in maize leaves exposed to osmotic stress, is a response to water deficit and allows the reduction of growth rates. The use of 2-D gels made it possible to detect the protein changes in response to water stress.

However, over the next few years, proteomics will help to overcome many of the most challenging problems in contemporary biology and biotechnology. Proteomics offers immense prospects in crop improvement revealing nature's secrets.

6

Gene Silencing for Biotic Stress in Plants

P.N. Jagadev

Gene silencing is the suppression of gene expression through nucleotide sequence-specific interactions that are mediated by RNA. Many manifestations of RNA silencing-RNA interference (RNAi) in animals, quelling in fungi and post-transcriptional gene silencing (PTGS) in plants, appear to be related. RNA silencing is experimentally activated by double-stranded RNA (ds RNA) and is used as a powerful technique for specific inhibition of gene expression in a variety of organisms. In a natural context, ds RNA may be produced from rearranged loci by transcription from converging promoters or by host-or viral-encoded RNA-dependent RNA polymerases (RdRP). A key component of RNA silencing is 21-23 nt RNA known as short interfering RNA (Si RNA). In Drosophila, the siRNA is derived from ds RNA by the action of an RNase-III like enzyme named DICER. The siRNA guides a multi-subunit endnuclease, referred to as the RNA-induced silencing complex (RISC) and it specifically degrades RNA that share sequence similarly with the ds RNA.

Much attention has been paid recently to RNAi, a technique in which exogenous, ds RNAs are introduced into a cell to specifically destroy a particular mRNA, thereby diminishing or abolishing gene expression. The technique has proven effective in *Drosophila*, *Caenorhabditis elegans* (a nematode), plants and recently, in mammalian cell cultures.

Since the early days of molecular biology, RNA has taken a back seat to DNA. RNA's status was upgraded in the early 1990s when Prof. Victor Ambros at Darmouth College found RNA strands in *C. elegans* that interfered with the nematode's mRNA. In 1999, when Prof. Andrew Fire at the Carnegie Institution of Washington and Prof. Craig Mello of University of Massachusetts discovered that ds RNA administered to nematodes could effectively turn off genes, the era of RNAi was born.

RNA silencing was first discovered in transgenic plants, where it was termed cosuppression or posttranscriptional gene silencing (PTGS). Sequence-specific RNA degradation processes related to PTGS have also been found in ciliates, fungi (quelling), and a variety of animals from *Caenorhabditis elegans* to mice (RNA interference RNAs (siRNAs). A key feature uniting the RNA silencing pathways in different organisms is the importance of double stranded RNA as a trigger or an intermediate.

RNA Silencing as a Defense Against Viruses

Several lines of research indicate that RNA silencing is a general antiviral defense mechanism in plants. The first indication came from studies of pathogen-derived resistance (PDR) in plants. In PDR, resistance to a particular virus is engineered by stably transforming plants with a transgene derived from the virus. Eventually, it became clear that one class of PDR was the result of RNA silencing of the viral transgene. Once RNA silencing of the transgene had been established, all RNAs with homology to the transge were degraded, including those derived from an infecting virus. Thus, plant viruses could be the target of RNA silencing induced by a transgene. The same work demonstrated that plant viruses could also induce RNA silencing. Virus-induced gene silencing (VIGS) can be targeted to either transgenes or endogenous genes and the technique has been used to screen for gene function using libraries of endogenous sequences cloned into a viral vector.

The idea that RNA silencing is an antiviral defense pathway is strengthened by observations of natural plant-virus interactions. First, plants recover from certain plant viral infections, and the recovered plants are resistant to re-infection by the initial virus (and to closely related viruses) because of an RNA silencing mechanism. Second, many plant viruses encode proteins that suppress RNA silencing, suggesting a co-evolution of defense and counter defense between the host and the invading viruses. These viral suppressions target different steps in the silencing pathway and have provided a new approach to understand the mechanism of RNA silencing in plants. It remains to be seen if viruses of fungi or animals use a similar counter defensive strategy against RNA silencing, or indeed, if RNA silencing serves as a defense against viruses in organisms other than plants.

In plants, RNA silencing can be induced locally and then spread throughout the organism and this aspect of the process likely reflects its role in viral defense. Plant viruses generally enter a cell at a small wound, replicate within that cell, and then move cell-to-cell until they reach the vascular tissue, which serves as a conduit to all parts of the plant body. The movement of the mobile silencing signal in the plant parallels that of the virus, traveling in the vascular tissue and spreading out from the veins. Thus, an invading virus enters a race with the host. If the virus moves faster, it can establish a systemic infection. If the silencing signal goes faster, then the virus will enter systemic tissues only to find RNA silencing already established and the infection will be aborted. Given the defensive nature of the mobile silencing signal, it is not surprising that some viruses encode proteins that interfere with the production or movement of the signal. The mechanism for systemic spread of silencing is one of the big unanswered questions in the field and also one of the most difficult to address experimentally. It may well be that plant viral suppressions will provide a powerful set of tools to examine this aspect of silencing.

Recent studies of two plant viral suppressors, the helper component proteinase (HC-Pro) of potyviruses and the p25 protein encoded by potato virus X(PVX), represent two different viral strategies to suppress silencing. HC-Pro is a highly effective suppressor of silencing that can enhance the accumulation of a broad range of unrelated plant viruses, likely accounting for the large number of potyvirus-associated synergistic diseases in plants. It prevents both VIGS and transgene-induced RNA silencing is reversed at a step that eliminates the accumulation of the SiRNAs, but fails to eliminate the mobile silencing signal as assayed by grafting experiments. In contrast, PVX p25 is much less effective in blocking silencing than HC-Pro and it appears to target and interfere with systemic silencing.

Conclusion

Gene silencing is an important genetic mechanism with both theoretical and practical implications. The discovery of a viral sequence that suppresses gene silencing opens a new and promising avenue to understanding the mechanisms that regulate the pathway in plants. Previously, it was found that transgenic plants expressing the 5'-proximal region of the tobacco etch potyviral (TEV) genome (termed the P1/HC.-Pro sequence) develop synergistic disease when infected with any of a broad range of plant viruses. The expression of the P1/ HC–Pro sequence might interfere with a general antiviral system in plants, thereby permitting viruses to accumulate beyond the normal host mediated limits. P1/ HC-Pro can be used to help establish whether different classes of gene silencing employ similar mechanisms of pathways. The practical implication of the finding centers on potential improvements in genetic engineering in plants, where it is desirable to manipulate silencing. Technologies that use plants to express foreign gene products or to overexpress endangered genes are often hampered by gene silencing, which prevents long term high level expression of the transgene. It may be possible to use the P1/HC-Pro silencing suppressor to directly counter silencing in these situations. Elucidation of the mechanism by which P1/HC-Pro suppresses silencing may also lead to the development of strategies to enhance silencing in plants, thereby improving technologies that exploit gene silencing to eliminate expression of undesirable genes.

7

Biotechnology in Horticulture

Sashikala Beura

Biotechnology provides significant opportunities and greater effects in horticulture, where even minor changes such as quality, colour, aroma, flavour, structure, disease-pest resistance and postharvest behaviour bring tremendous achievements commercially. Tissue culture, genetic engineering, molecular breeding, biofertilizers, biopesticides and postharvest biotechnology are important aspects of biotechnology in horticulture such as fruit, vegetable, ornamental, spice, medicinal, aromatic and plantation crops.

Tissue Culture

Tissue culture is the true reality of the plant biotechnology. The annual demand of tissue cultured production constitutes nearly 10 per cent of the total world biotech business, amounting to 15 billion US dollars. The global production of tissue culture industry is expanding @ 15% per annum (Chadha, 2000). In India, more than 130 tissue culture units are presently producing > 300 million plantlets per annum (Choudhury and Tejaswini, 2000).

Department of Biotechnology, Government of India has set up pilot-scale facilities including Micropropagation Technology Parks (MTPS) at the Tata Energy Research Institute(TERI), New Delhi; National Chemical Laboratory (NCL), Pune and National Facility for Virus Diognostics and Quality Control of Tissue Cultured Raised Plants (NFVDQTP) at IARI, New Delhi, with 5 satellite centres at NCL, Pune; TERI, New Delhi; IHBT, Palampur: IIHR, Bangalore and SPIC, Chennai (Singh et al., 2004).

Micropropagation

Micropropagation is the most effective method for production of true-to-type disease free and healthy plantlets in horticultural crops. Micropropagation is a reality in fruit crops like banana and strawberry and to some extent, success has been achieved in pineapple, grape and papaya. In apple and citrus, protocols for regeneration of root stocks have been developed. However, it is the need of hour to develop protocols for multiplication of sapota, guava, litchi, coconut, oil palm and datepalm.

In vegetables, viable protocols for micropropagation have been used in potato, cassava, sweet potato, asparagus, cauliflower, carrot, garlic, ginger and turmeric etc. In floriculture, *in vitro* method of propagation is used commercially in gerbera, anthurium, carnation, mussaenda, orchid, rose, gladiolus, dracaena, *Aglaonema* and *Philodendron* etc. The technology is advantageous for those crops, which can not be multiplied rapidly by conventional method as in case of datepalm and coconut etc.

Embryo Rescue and Culture

The method of embryo rescue and its culture on artificial medium is an effective tool for combating the problems of post-zygotic failure of embryos. Success has been achieved in hybrids of *Musa acuminata* x *M. bulbisiana*, *Carica papaya* x *C. cauliflora* interespecific crosses in pineapple, seedless grape, citrus and tomato. Intergeneric hybrids include *Raphanus sativus* x *Brassica napus*. Hybrid raised by embryo culture of *Lycopersicon esculentum* x *L. peruvianum* is resistant to virus, fungi and nematodes and *Solanum melongena* x *S. khasianum* is resistant to brinjal shoot and fruit borer, where as *Solanum tuberosum* x *S. etuberosum* is resistant to potato leaf roll virus.

Syn Seed Technology

Different micropropagating materials such as somatic embryos, shoot buds, shoot tips, embryogenic calli formed *in vitro* are coated (encapsulated) in gel containing nutrients and other additives and designated as synthetic or artificial seeds. Successfully, syn seeds are produced in mulberry, mango, banana, potato, carrot, brinjal, *Dioscorea* and orchids (Beura, 2003). Artificial seeds can be stored, transported and ultimately made to germinate in culture (*in vitro*) or directly on soil (*in vivo*).

Germplasm Storage and Cryopreservation

Crop breeding programme largely depends on the preservation of germplasm to ensure the availability of useful germplasm at anytime. Storage of meristem/ somatic embryos at ultra-low temperature in liquid nitrogen (-196 °C) and subsequent plant retrieval is known as cryopreservation. Among various horticultural crops, somatic embryos and pollen embryos of *Citrus* spp., cell suspension and callus of *Capsicum* spp. somatic embryos of *Coffea arabica* and *Daucus carota*, meristems of *Solanum tubersoum* are suitably cryopreserved for longer period.

Genetic Engineering

This technique introduces a specific gene of interest from a donor species into the genome of a host species. The methods for genetic transformation include inoculation of explants with *Agrobacterium tumefaciens* or *A. rhizogenes*, bombardment of apical meristem followed by co-cultivation with *A. tumefaciens*, microprojectile bombardment of cell with DNA coated particles and protoplast electroporation. The expression of undesirable genes can be blocked by the application of antisense technology.

Disease Resistance

The coat protein mediated approach to engineer virus resistance has been applied to introduce resistance against disease like Pul M Pox Virus (PPV), Papaya Ring Spot Virus (PRSV), Citrus Tristeza Virus (CTV), Grape Fan Leaf Virus (GFLV) PVX and PVY in potato and CEV d in tomato.Some success has been achieved in fungal and bacterial disease resistance of apple, potato, tobacco, tomato, carrot and anthurium.

Insect Resistance

The bioinsecticidal -endotoxin gene (Bt gene) isolated from *Bacillus thuringiensis* is currently used commercially in plants to make them resistant to insect pests. The crystal protein has shown a good response against the lepidopterans, coleopterans and dipterans. The success has been achieved in potato, tomato, brinjal, cabbage, broccoli, Chinese cabbage and rutabaga.

Transgenics for Quality

The first transgenic crop for commercial sale was transgenic tomato "*Flavr Savr*' with delayed ripening, developed by Calgene, USA in 1994. Genetically modified vegetables such as broccoli with prolonged shelf life, carrot, potato and cassava with improved nutrition, melon, pepper and tomato with delayed ripening, sweet potato with altered amino acid composition and improved protein are now available.

Flower Colour

The first application of genetic engineering was to modify the flower colour in petunia. Orange coloured petunia variety was developed.This was achieved by expressing the dihydro.-flavonol-4-reductase gene 'dfr' from maize in a suitable petunia line. Then altered flower colour chrysanthemum was developed. Since ornamentals are not grown as a food source, consumers may have readily accept genetically engineered ornamental crops than genetically engineered food crops.

Vase Life

Many cut flowers deteriorate rapidly after harvesting, and short vase life has been a major obstacle in their marketing because of production of ethylene. Transgenic carnations such as 'Scania' and 'White Sim' cultivars were produced with an antisense acogene. Their flowers produced low level of ethylene, thus increased the vase life.

Fragrance

Monoterpenes are an important class of floral fragrance compounds and are used for perfume production. They are derived from geranyl pyrophosphate by various monoterpene synthases. Recently one monoterpene synthase (S-linalool synthase) has been cloned. Two genes that modify monoterpenes have also been cloned recently. There are no reports of the modification of floral fragrances using genetic transformation, but isolation of monoterpene synthases and modification of genes makes such work possible in future.

Molecular marker analysis has been done in various ornamental plants and DNA fingerprinting has been standardized in rose, crysanthemum and *Rhododendron* for accurate identification of plant cultivars.

Conclusion

Plant biotechnology has opened new avenues and opportunities, especially tissue culture and genetic engineering in horticulture to combat all kind of challenges. Transfer of desirable genes irrespective of cultivars, species, genera and even taxa is now possible through the wonderful techniques of genetic engineering. In future, transgenic male sterility for hybrid seed production of vegetables, parthenocarpy for seedlessness in brinjal, cucumber and guava, enhancing antioxidants in vegetables, development of blue colour rose and scented gladiolus are the most awaiting dreams of the horticulturists to become the reality. Thus, farmers and consumers will be benefitted from the genetically modified fruits, vegetables, especially ornamental plants where they do not enter directly into food chain.

8

Multivariate and Numerical Taxonomic Analyses for Agricultural Productivity

L.D. Mishra and B. Pradhan

The success of any crop improvement programme springs mainly from sound and genetically investigations carried out in the fields. However, the moment such biological populations are subjected to field evaluation, some external and non-heritable agencies come into active role. As a result, these agencies or factors mask real differences among the tested populations which stem purely from heritable or genetic factors. Their influence gets confounded with the latter. Consequently authentic data related to genetic differences are not available due to controllable and non-controllable errors. Further, physical data that are assembled in an experiment are often so voluminous and so vast that is perhaps humanly not possible to arrive at meaningful conclusions. Hence, an adequate knowledge of statistical principles and their appropriate use warrant perfection to a point where all possible sources of error can be well accounted for.

Multivariate Analysis

Conservation of genetic resources is the basic step to crop improvement programme for which grouping of genetic stocks based on suitable scales and choice of genetically divergent parents under transgressive breeding programme is quite imperative and essential. The systematic classification mainly depends upon easily distinguishable qualitative attributes which represents phylogenetic hierarchy of individuals or group of individuals. But it does not consider quantitative traits showing continuous variation which constitute the backbone of plant breeding practices.

The quantitative classification offers a quantified degree of divergence among genotypes or populations and serves as a sound basis of grouping any two or more genotypes based on minimum divergence or resemblance between them. There are different models of multivariate analysis, used in crop improvement programme. They are as follows:

D2- Analysis: it gives quantitative measurement of divergence among the genotypes.

Canonical Analysis: serves in terms of spatial distance Duncan's Multiple Range Test (DMRT) Metroglyph Analysis

Index Score Analysis

Factor Analysis

D2 - Analysis

For the first time it was proposed by Prof. P.C. Mahalanobis (1936). It determines the divergence among populations in terms of generalized group distance. It has been successfully used by Rao, 1952 and Murty and Arunachalam (1966) in different types of classification studies. In this method the D2 values are estimated by using the following formula.

$$D_x^2 \sum P_i \sum P_j (X_{ij}) d_{ij}$$

Where, X = number of metric traits

P = number of genotypes

^{1}ij = matrix reciprocal to common dispersion matrix

d_i and d_j = difference between the mean values of two genotypes for i_{th} and j_{th} characters respectively

Thus, D2 = sum of squares of difference between any two genotypes for each of the uncorrelated variables.

Computational procedure

The entire computational procedure involves the following steps.

1. Collection of replication wise mean data
2. Test of significance of differences among the genotypes
3. Transformation of correlated variables (XS) into uncorrelated (YS) variables
4. Comparison of D^2 values
5. Test of significance of D^2 values
6. Group constellation
7. Computation of intra-and inter-cluster divergence

For group constellation, the limit of divergence within groups should be based upon the range of all $\frac{n(n-1)}{2}D^2$ values and the divergence between groups 2 (inter-cluster) distance should not be lower than that within groups (intra-cluster average D^2 values).

Interpretation and implication of D^2 values

The range of D2 values indicates the degree of divergence among the tested genotypes. Intra-cluster distances indicate the divergence between the genotypes included in a single cluster while inter-cluster distances indicate the divergence between the clusters. Cluster means indicate average performance of all varieties clubbed in a cluster. Finally choice of divergent parent(s) can be made for hybridization purposes on the basis of D2 values between two genotypes or two clusters.

Limitations

It is very lengthy and cumbersome, with increase in number of genotypes problem multiplies many fold. Since the technique is based on second order statistics, the estimates are not robust and can not be applied to non-replicated data.

Advantages

D2-analysis helps in selection of genetically divergent parents for their exploitation in hybridization programme, measures the degree of diversification and determines the relative proportion of each component character to the total divergence.

Canonical Analysis

It is a sort of multivariate analysis where canonical vectors and roots representing different axes of differentiation and the amount of variation accounted for by each of such axes respectively, are derived. It is also called as "Vector Analysis" or "Principal Component Analysis". Most of the steps of computational procedure is similar to D2-analysis. The subsequent steps to D2-analysis are derivation of canonical vectors 1 and 2 and canonical roots 1 and 2. The Canonical Analysis measures the divergence between genotypes in terms of spatial distance rather then quantifying it as D2 does. Thus it confirms the group-constellation obtained by D2-analysis.

Metroglyph Analysis

It is a semigraphic method of studying variability in a large number of germplasm lines taken at a time. It was for the first time proposed by Anderson for classification of crop plants e.g. cotton, pulses and oilseeds.

Assessment of variability by semigraphic method (metroglyph) of Anderson (1957) involves plotting of glyph i.e. a small circle by which the position of a genotype or line is represented on the graphs is carried out. For plotting of glyph-two characters with high variability are chosen (one in X-axis and the second in Y-axis). The exotic and indigenous lines may be represented by solid and open glyphs. The variation for remaining characters of each genotype is displayed on the respective glyph by rays. Each character occupies a definite ray position (vertical, inclined to right and inclined to left) and variation for each character is depicted by the length of ray (short, medium and long).

Index Score Analysis

In index score analysis the variation for each character is divided into three groups (low, medium and high) and the genotypes with low, medium and high values are given index score 1, 2 and 3 respectively. Finally the worth of a genotype is calculated by adding the index values of all the characters.

The genotypes can also be grouped in different groups or clusters by dividing the X and Y axes into three groups (low, medium and high). Thus, the maximum number of groups or clusters will be nine. Finally, the variation can be analyzed for various traits within the group and between the groups. It is a semi-quantitative technique useful for grouping the germplasm lines at a time. The procedure is very simple and can be applied to both replicated as well as non-replicated data as the analysis is based on the mean values of each character. On the contrary, inclusion of large number of genotypes and study of more number of characters at a time is difficult.

Numerical Taxonomy (NT)

The grouping of taxonomic units by numerical method into different taxa on the basis of their characters refers to "Numerical Taxonomy". Through NT approach we draw phylogenetic inferences from the data by statistical/other mathematical methods. Numerical taxonomic method is preferred to quantitative taxonomy since the latter does not include method using serology or paper chromatography data.

This method requires the conversion of information about taxonomic entities into numerical quantities. Every character is given equal weightage in creating natural tax. The classification is based on phenetic similarity.

The greater the content of information in the taxa of a classification and the ore characters on which it is based, the better will be classification. The overall similarity between any two entities is a function of their individual similarities. The phylogenetic inferences can be made from the taxonomic structures of a group and from character correlation. The most important and fundamental step in NT is the collection of information about many characters. All kinds of characters (morphological, physiological, ethnological etc.) are equally desirable.

The classification in NT is generally based on a matrix of resemblance and taxa are constructed through various techniques. In numerical classificatory analysis, the 'general similarity coefficient' of Gower (S_G) is used as a measure of resemblance between different operational taxonomic units or OUT's. The S_G values can be calculated following Sneath and Sokal (1973). Basing on the matrix of S_G values, phenograms (dendrograms) are constructed using UPGMA (Unweighted pair group method using arithmetic average) clustering strategy of SAHA (Sequential, agglomerative, hierarchic, non-overlapping) clustering method. Finally clusters are identified at appropriate phenon levels.

References

Anderson, T.W. 1957. An introduction of multivariate statistical analysis. John Willey and Sons, New York.

Mahalanobis, P.C. 1936. On the generalized distance in statistics. *Proc. Nat. Sci* (India), **12**: 49-55. Rao, C.R. 1952. Advanced statistical methods in biometrical research. 1st Edn. John Willey and Sons, New York.

Murty, B.R. and Arunachalam, V. 1966. The nature of genetic divergence in relation to breeding system in some crop plants. *Indian J. Genet.,* **26** A: 188-198.

Sneath, P.H.A and Sokal, R.R. 1973. "Numerical Taxonomy". W.H. Freeman and Co., San Francisco.

9

Role of Bioinformatics in Plant Pathological Applications

N.K. Dhal

The importance of Plant Pathology has been realized over the years owing to number of crop failures in different parts of the world. Plant disease control is indispensable for qualitative production and to compete in global agricultural trade. There is great need of sensitization of our research in various aspects which shall guide the generation to come for managing plant diseases not only for enhancing crop production but also in facilitating trade by conserving the biodiversity and environment.

Research Trend in Plant Pathology

Like in other sciences, there has been a change in the trend of research in Plant Pathology with passage of time. Up to the late 1960s most of the interest in plant pathology was concentrated in crop diseases mostly caused by fungi. Physiology of microorganisms and physiology of parasitism became increasingly popular fields of interest in 1960s.

Among disease control aspects, screening for host resistance and chemical methods of disease control have steadily shown considerable increase in attention since the late 1960s. Another area currently of rapid growth of interest is biochemical plant pathology concerned with physiology of parasitism and nature of host resistance.

There have been several new areas of interest since 1970s. These include genetics of microorganism, loss assessment, seed pathology, mycotoxicology, molecular plant pathology. Research efforts also intensified in area like virus diseases, plant disease appraisal, epidemiology, economics of chemical control and integrated control strategy.

Scope for Application of Bioinformatics Tools

Disease diagnosis and management

One of the great break through in the field of plant pathology has revolutionalized agricultural sector was discovery and introduction of different chemical for control of plant diseases. However, disease diagnosis is based on symptom development and causal organism are the prerequisite for successful management of diseases. It is always not possible to recognize the disease at its proper stage of development for control of the disease. Further, control of plant diseases with suitable chemical is a flexible method that can be scheduled well in advance as prophylactic measure to prevent development of disease as therapeutic measures. Therefore, bioinformatics tools can be used for quick identification of the disease and advise the farmers on various aspect of its management.

Identification and detection of plant pathogens

A large number of different microorganisms are responsible for causing plant diseases. Researchers find it different to identify the organism to oven up to the species level. Computer assisted diagnostic tools can be used for identification of plant pathogens.

Disease forecasting

Occurrence and spread of a disease depend on weather parameters and inoculums load besides the host and pathogen factors. Plant disease forecasting is considered as an important area of research for economic management of diseases. Disease forecasting models can be developed taking into consideration of various factors of disease development using bio-informatic tools.

Molecular plant pathology

Molecular plant pathology is one of the fastest moving and most exciting fields in plant pathology which is directly benefited from advance in molecular biology and recombinant DNA technology. These have been applied to both pathogen and plants, allowing to develop understanding of organisms within themselves and the complex plant microbe interaction leading to compatibility and incompatibility. Two areas of research that have received major thrust in plant pathology and need for use of bioinformatics tools are:

a. Search for molecular markers, cloning disease resistance genes and understanding signal transduction mechanism for disease resistance.

b. Development of molecular diagnostic tools for pathogen detection, determination of genetic diversity among plant pathogens, strains and their polygenetic relationship.

Conclusion

Plant pathology played significant role in agricultural development of the country. Nevertheless, new challenges to our economy are unending. Although we have developed tremendous expertise in conventional plant pathology research, we must strive hard to intensity quality research and application using bioinformatic tools.

10

Informatics Tools for Conservation and Utilization of Biological Resources with Special Reference to Agro Biodiversity

P. K. Panda

The focus of world's trade and economic activities is increasingly being shifted from resource-based economics to knowledge-based economics. This paradigm shift has been triggered due to the implementation of Convention on Biodiversity (CBD) and other related international treaties and policy frameworks. CBD for the first time states that bio-resources are the sovereign property of the countries in which they are found. It further stipulates that access to bio-resources by outside parties can be made only after prior approved consent of the concerned stakeholders and also with mutually agreed terms and conditions for equitable sharing of benefits accrued from the commercialization of biodiversity. To realize such proclaimed rights given in CBD and to fulfill obligations of various other relevant treaties, particularly in the context WTO regulation, it is imperative that the bio-resources-rich nations like India have comprehensive knowledgebase of the biodiversity and associated information systems in accessible form.

India ranks among the top twelve nations of the world in terms of biological diversity and is also one of the early centers of crop domestication which includes such important ones as rice, sugarcane, banana, citrus, mango, eggplant, black pepper, neem etc. The ingenuity of traditional farming communities resulting in identification and adoption of many non-traditional and un-explored economic plants have further contributed to the country's agro-biodiversity. These genetic resources and land races not only represent the invaluable source for novel genes for future crop improvement programmes, but also provide local food security to resource-poor marginal farmers, without compromising on sustainability of agro-ecosystems.

Sustainable management of invaluable bio-wealth depends upon effective integration of multi-disciplinary and complex biodiversity datasets by using the emerging bioinformatics tools. Biodiversity informatics harnesses the power of computational and information technologies to organize and analyze data on plants and animals from genomic to ecosystem levels. Relevant expertise and synergy of multi-disciplinary researchers and computation experts are required to develop such multi-faceted databases.

11

Applications of Bioinformatics in Aquaculture

A.K. Roy and N. Sarangi

Bioinformatics is a discipline that deals with computation of biological information. Bioinformatics is a newly emerging interdisciplinary research area spanning a range of specialties that include biology, biophysics, computer science, mathematics & statistics. It makes use of scientific and technological advances in the area of computer science, information technology and communication technology to solve complex problems in life sciences particularly problem in Biotechnology.

Explosive growth in biological data from large sequencing projects that are producing nucleotide sequences continuously at a faster rate and the content of the nucleotide database is doubling in every 14 months. The latest release of Gene bank (v.102) exceeded one billion base pairs. To cope with this huge volume of data, a new scientific discipline has emerged comprising bioinformatics, biocomputing & computational biology.

Bioinformatics Centre on Aquaculture: The Bioinformatics Centre on Aquaculture established at Central Institute of Freshwater Aquaculture (ICAR), Kausalyaganga, Bhubaneswar is a Distributed Information Sub-Centre (Sub-DIC) under Biotechnology Information System (BTIS) Network of the Department of Biotechnology, Government of India during 1991-92. The centre is specialized in the field of aquaculture and serves as an information source in the country. The objectives of the centre is to function as an information base in the field of aquaculture in the country so that scientists working in the field get ready access to the computer based information on resources, databases, besides providing computer based workshop and storage and retrieval system, extending communication linkage, to develop software packages, to conduct training programs to develop human resources and to create awareness about computerized storage and retrieval system among researchers, planners, developers working in the field of aquaculture.

Hardware: Informatics-based infrastructure includes the latest computers and peripherals, network and communication systems. Databases were procured, installed and put to use in the Bioinformatics Centre of CIFA. The Hardware are: P-4 Multimedia Computers, 1 & 3 KVA online UPS, Dot Matrix/ DeskJet/ HP LaserJet 1320 N (Network) Printers, LCD Projector, Colour Scanner, Modem, CD Writer, 5.5 KVA Generator, Hubs, Switches and Routers.

Software: Procured MS Office 97, 2000, Novel Netware 4.1, FoxPro, QPRO, Oracle, Visual Basic, M.S. Project, Visual Studio. The Operating Systems are: DOS, Windows 95, 98, UNIX, Linux.

Scientific Software packages: Quantity one (gel analysis), DNA Star, Reference Manager, UVI – Software (gel analysis)

Statistical Packages: The Institute has procured the latest version of Statistical Package SPSS (Statistical Package for Social Sciences, Version 14.0) and SAS (Statistical Analysis Systems) for computational work and exploratory statistical analysis

Databases Procured: Acquired the following CD-ROM databases/diskettes related to fisheries/aquaculture from various international organizations engaged in fisheries research and development to facilitate offline retrieval of literature search. (1) Aquatic Sciences and Fisheries Abstracts (ASFA) 1978 to June, 2003 produced by Cambridge Scientific Abstracts (CSA). ASFA On-line is also subscribed from 1971 to current period with IP based service connectivity provided to scientists and research workers of the Institute. (2) Fish Base, 1997; (3) The FAO Fisheries Atlas, 2001 – Preliminary Release – FAO, Rome, Italy, (4) Diagnosis of Shrimp Disease, 2000 – V. FAO, Rome and NACA, Thailand. (5) Maxims, Ecopath and Fish growth parameters software packages in diskettes, ICLARM, Manila, (6) FAO - ICLARM (Stock assessment tool), (7) FishstatPlus (V 2.3) – FAO, Rome, 2000, (8) Archives of the Bay of Bengal Programme 1979-2000.

Communication and Network Linkage: Installed and operating Leased Line with 512 kbps bandwidth from BSNL for high speed Internet & browsing and e-mail uploading and downloading. The centre has also acquired installed and operating VSAT sky blaster with 256 kbps bandwidth for Internet surfing. SLIP/ PPP Connectivity from NIC for Internet browsing is also maintained.

Local Area Networking (LAN): More than one hundred computers are distributed to different divisions/sections of the Institute and connected with LAN for sharing of information resources and 42 nodes are connected with high speed Internet browsing facility.

Identification of Fish Related Databases (Online):

(i) A global information system on fishes (http://www.fishbase.org)

(ii) Online information for medaka ricefish (www.bio1.bio.nagoyau.ac.jp. 8000)

(iii) Information on the pocciliid fish xiphophorus (www.xiphorus.org)

Web Publication for Knowledge Dissemination: Designed, developed content created and hosted the webpage of CIFA of about 260 MB size covering various aspects like Profile, Director's Office, Staff, Division, Research Projects, Collaborations, Technologies, Research Contributions, Transfer of Technology, Publications, Consultancy, BTIS profile, Training Programme, Statistics, Products, Recognition, About Orissa, News letters, Annual reports and Useful Links to facilitate free flow of Information and sharing of knowledge countrywide. [http://www.cifa.in]

Library Information System: The library system is computerized. CDS/ISIS package is used for the author-wise/ title-wise/discipline-wise search of books/ journals available in the library.

Exploratory Data Analysis

Development of Biophysical Models for Catla catla, L. rohita, H. molitrix, C. idella and C. mrigala: Aquaculture is considered as a black box. A number of factors interact dynamically and influence fish growth and production. Therefore, attempt was taken to identify and quantify various factors responsible for the growth of different species of carps reared in sewage fed ponds and to develop models. Multivariable analysis was performed with Specific Growth Rate (SGR) as dependent variable and temperature, available phosphorus, plankton, pH, chloride, free CO_2, BOD, alkalinity, NH_4-N, DO, Initial weight as independent variables using various regression procedures. Details of variables considered for the study, regression methods, co-efficient of determination (R2) values and variables selected in the regression equation along with contribution of each of the selected biotic and abiotic factors for different species are presented in the table below (Roy *et al.*, 2000).

Table 1 : Salient features of models developed for different species

Species	Variable considered for study	Regression Methods	Co-efficient of determination (R^2)%	Contribution of independent variables selected in the regression equations
Labeo rohita	Av. Phosphorus Initial Weight Temperature Alkalinity Free CO_2 Chloride Plankton NH_4-N BOD DO pH	Backward/ Forward	80.96	Av. P DO INIT WT PL Free CO_2
Catla catla	-do-	-do-	87.93	Initial weight, temperature, benthos, plankton, P_2O_5, DO and BOD
H. molitrix	-do-	-do-	72.14	DO Av. P NH_4-N
C. idella	-do-	-do-	70.58	DO Av. P
C. mirgala	-do- + Benthos	Forward Backward	69.45 74.89	DO Av. P NH_4-N Av. P, DO, NH_4, Temp.

Source: Anon, 2006 & Roy *et al*., 2000

Factor analysis & Path analysis: Factor analysis was performed on the data set of sewage-fed fish culture pertaining to the growth of *Catla catla* in order to bring down the factors capable of explaining most of the variances. Kaiser's Criterion was used to select those factors which have an eigen value of greater than one. Graphical screen test was also performed to decide which factors are to be excluded. *Product moment correlation coefficient* was calculated which is the indicator of degree of interrelationship that exist within the data. Secondly initial statistics of each variable was worked out to find out variance explained by each factor. It is observed that linear combination formed by factor 1 has a variance of 3.396 which is 30.9% of total variance. Similarly second, third and fourth factor explain 19.2, 17.5 and 11.3% of total variability. Thirdly from factor matrix which contains coefficients that relate the variables to the four factors was worked out in the final statistics, proportion of variance explained by the factors known as communality of variable was calculated along with eigen values and percentage of variance explained by the factors and lastly from the rotated factor matrix.

Table 2: Rotated Factor Matrix showing factor loading for different variables related to growth of *Catla catla* under sewage-fed culture system

	Factor 1	Factor 2	Factor 3	Factor 4
pH	0.90788	0.10270	0.09692	0.26583
Alkalinity	0.82717	-0.20318	0.13040	-0.04890
BOD	0.80173	-0.25360	-0.07303	-0.10913
Growth/day	0.02612	0.82241	-0.31062	-0.07163
Init.Wt.	0.26204	-0.70575	-0.23398	0.30780
DO	-0.49910	0.67964	0.16221	0.27600
NH_4-N	0.07116	-0.12673	0.88226	-0.02631
LNB	-0.12824	0.32691	0.67337	0.39762
Av. P.	0.53060	-0.06920	0.67083	-0.34344
LNPL	-0.02747	0.07424	0.14865	0.85969
Chloride	0.06325	-0.38773	-0.20215	0.79829

Source: Anon, 2006

It is evident that pH, alkalinity and BOD which are indicators of pollution load have higher factor loading for factor 1 which is crucial for sewage-fed ecosystem. Increment per day, initial weight and DO have higher factor loading for factor 2 representing two biotic factors. NH_4, available phosphorus and benthos have higher factor loading for factor 3 where first two factors represent nutrient status of sewage-fed ecosystem and finally chloride and plankton have higher factor loading for factor 4 where plankton plays a major role in the growth of catla in pond ecosystem.

Path analysis was done to identify and visualize the direct and indirect effect of various biotic and abiotic factors on the Specific Growth Rate (SGR) of silver carp. Variables considered for the study are DO, BOD, pH, P_2O_5, NH_4-N, Alkalinity, Chloride, Free CO_2, Temperature, Plankton, Benthos and Initial weight of the test species reared in sewage fed pond of Wastewater Aquaculture Division of CIFA. Significant positive correlation was observed between SGR and each of the variables like DO, Av. P., plankton, pH and temperature. SGR showed negative correlation with each of the free CO_2 and initial weight of the test species. Taking significant contributory variables in the path equation further analysis was done which shows the direct and indirect effects that are presented in the correlation matrix (Direct effects on main diagonal): (Table 3).

Table 3: Matrix of direct and indirect effects

	Initial wt	DO	Av. P.	LnPl	CO_2	Temp.
Initial wt	-0.4815	-0.1283	-0.0758	0.0348	0.0907	-0.0018
DO	0.1695	0.3643	0.0461	0.0944	-0.0541	0.0057
Av. P.	0.1414	0.0650	0.2581	0.0753	-0.0222	0.0006
LnPl	-0.0570	0.1169	0.0661	0.2943	-0.0308	0.0013
CO2	-0.2522	-0.1138	-0.0331	-0.0524	0.1732	-0.0021
Temp	0.0453	0.1086	0.0076	0.0200	-0.0194	0.0190

Source: Anon, 2006

For the first time the concept of path analysis has been introduced in Aquaculture experimental data to visualize the effect of various biotic and abiotic factors on the growth of silver carp.

Statistical Analysis of 421 breeding experiments of *L. rohita* and salient findings

Data on breeding experiments (421 sets) of *L. rohita* conducted during 1970-1982 were computerized and analyzed statistically to study the breeding performance of these fishes in relation to biotic and abiotic factors. The treatments used for breeding were divided into the following broad heads. Carp Pituitary Extract (CPE) – H1: Non-carp Pituitary Extract (NCPE) – H2 and Gonadal Hormone (GCH) - H3.

The qualitative characters (treatment and response) were classified as follows showing the number of occurrences under each category along with percentages (Table 4).

Table 4: Breeding response of *L. rohita* under various hormone treatments

Treatments	Response		
	Positive response	Lack of response	Total
H1 (CPE)	274(82.30)	59(17.7)	333
H2 (NCPE)	12(28.6)	30(71.4)	42
H3 (GCH)	22(47.8)	24(52.2)	46
Total	308	113	421

Source: Gupta *et al.,* 1999; (Values in parenthesis indicate percentage)

Two-way classification of the above experimental results gave comparative performances of various treatments over response and X2 (chi-square) test of the data reveals that (X2 tab = 5.99<X2 cal = 48.45) response depends upon the treatments. **Analysis of Variance (ANOVA)** carried out to study the effect of hormone on egg release did not demonstrate any significant difference. However, maximum average egg release was observed in case of H1 (3.11)

followed by H2 (3.04) and H3 (3.0 lakhs). ANOVA performed to study the effect of hormone on spawn recovery did not reveal any significant difference. However, apparent highest spawn recovery was observed in case of H3 (1.53 lakhs) followed by H2 (1.22 lakhs) and H1 (1.09 lakhs). ANOVA revealed significant difference in percentage fertilization due to hormone at 5% level. Least significant difference value was calculated to be 14.95. Highest percentage fertilization was demonstrated by H3 (56.01).

Multivariate analysis of breeding data: In order to quantify the possible effect of different independent variables like weight of female, doses of hormone, temperature, quantity of eggs released and percentage fertilization on the dependent variable like spawn recovery a multivariate approach of data analysis was taken using SPSS package. The stepwise regression analysis was done for each of the hormone groups (viz. H1, H2 and H3) separately. The process involved calculation descriptive statistics followed by correlation matrix and finally backward regression procedure along with residual studies, testing of assumptions and model building. In case of Carp Pituitary Extract (H1) independent variables like female weight, temperature and quantity of eggs together explain 46.74% of the total variability in spawn recovery (R2 = 0.4674; $P<0.01$; n=212). Similarly in case of non-carp pituitary extract (H2) quantity of eggs alone explains 85.91% of total variability in spawn recovery (R2 = 0.8591%; $P<0.01$; n = 11). Further for gonadal hormone (H3) quantity of eggs demonstrated 21.33% of total variability (R2=0.2133; $P<0.05$; n = 16).

Identification of Aquaculture Related Information in the Digital Knowledge Bank

The impact of Internet on research & development activities of aquaculture sectors is tremendous in the present information era. The net offers tremendous potential for sharing resources like documents, data, software and network services. The Internet apart from being a vast storehouse of information is also a virtual meeting place. People can communicate through mailing lists, newsgroups, discussion forums, electronic bulletin boards etc. The range of information and expertise it offers includes: access to information in universities; organizations; and the accumulated wisdom of a number of professionals who share their experiences everyday with e-mail discussion groups. The Internet also helps sharing of information among individuals irrespective of location, replacing the traditional methods of exchange of information. In order to keep pace with new developments, Internet connectivity alone is not enough. What is needed is active participation and contribution to the world's growing digital knowledge bank that has the potential to educate and reform in many sectors including aquaculture. One way to facilitate this is through listservs or electronic discussion forums (Murthy, 2006).

Electronic Discussion Forums: Listservs are email discussion groups that are categorized by the area of interest. Discussion forums and listservs both offer opportunities to take part in online conversations with other professionals in the field. A Discussion Forum is an online area where contributors can carry on debates and discussions by posting questions and observations and inviting responses from other participants.

Listservs/Discussion Forums: Each week, the SeaWeb Aquaculture Clearinghouse releases a summary of aquaculture issues and developments from around the world by email. Email clearinghouse@seaweb.org with a quick note indicating your interest in subscribing to the Clearinghouse weekly online newsletter.

Aquanic: The Aquaculture Network Information Center (http://aquanic.org) is a gateway to the world's electronic aquaculture resources. The goals of AquaNIC are to provide access to all electronic aquaculture information at the national and international level; increase the quantity and quality of electronic information available to the aquaculture industry; provide self-paced aquaculture instruction to the aquaculture industry and obtain user input in directing AquaNIC services. Has a discussion group around common species and systems.

Aquaponics: (http://groups.yahoo.com/group/Aquaponics/) Aquaponics offers innovative methods of growing aquatic and terrestrial plants and animals. Drawing on and adapting technology, knowledge and experience from many fields including, intensive recirculating aquaculture, controlled environment, horticulture and hydroponics.

AquaBiz: The Aqua Business Club (http://groups.yahoo.com/group/ AquaBiz) is a online Aquarium/Aquaculture Business Club where, commercial aquarium fish breeders, home breeders, buyers, sellers, traders, importers, exporters, collectors, wholesalers, agents, trans-shippers, transporters, equipment dealers, live and dry food breeders/manufacturers/dealers can buy and sell their fish, invertebrates, related goods and services.

Aqua-culture (http://groups.yahoo.com/group/Aqua-culture/) has currently many members discussing and sharing information on all aspects of aquaculture.

Aquaculture (http://finance.groups.yahoo.com/group/aquaculture/) is an e-group formed for the benefit of aquaculturists, scientists, seafood traders, equipment manufacturers, aquaculture product manufacturers and marketers.

Network of Aquaculture Centres in Asia-Pacific–Forums (http://www.enaca.org/modules/newbb/index.php) NACA is dedicated to rural development and poverty alleviation through sustainable aquaculture and aquatic resource management.

Water-Forum: A moderated discussion forum for water resources issues; including drinking water, water conservation, environmental chemistry, wetlands, wastewater, irrigation, recreational use, fisheries and wildlife use, aquaculture, coastal studies and oceanography, environmental and public health issues, contamination/remediation, computer modeling, climatology, hydropower, and any other relevant water resources topics. To subscribe, one may mail to waterforum- subscribe@egroups.com

Aquq-L: An aquaculture discussion group owned by the Aquaculture Association of Canada and maintained by the Fisheries and Marine Institute of Memorial University of Newfoundland. To Subscribe to AQUA-L: Send a message to majordomo@killick.mi.mun.ca

Fishfolk: Fisheries social science network. Fish folk at MIT is a listserve devoted to fisheries issues of relevance to social scientists, fishermen, biologists, managers, government officials, conservation group members, attorneys, and other academics. Through discussions, members communicate lessons learned in one part of the globe that can help instruct others, and aid in sustaining the world's fisheries. Mail to listserv@mitvma.mit.edu to subscribe to the list.

Fish-Junior: To subscribe send an e-mail to listserv@searn.sunet.se with the message SUBSCRIBE FISH-JUN Your Name.

OCEANF-L: Ocean Farmers of America Forum. OCEANF-L is a discussion list to support the open exchange of ideas and information on:

1. The engineering of ocean farming systems for survivability, economy, and improved productivity,
2. Mechanisms to encourage fishermen to expand into ocean farming,
3. Biological, husbandry, and environmental issues associated with ocean farming at exposed sites,
4. Regulation and ownership of ocean farms in the U.S. Exclusive Economic Zone, and
5. Livestock registration and ownership of biological inventory Mail to listserv@mitvma.mit.edu to subscribe.

Taking Advantage of Virtual Library on Knowledge Management Network

World's most reputed knowledge management resource and it holds award winning content, community, research and practices (http://www.brint.com/km)

International Collaboration

The Centre collaborates with various International Organizations like FAO, UNDP, NACA and World Fish Centre for exchange of information resources related to fisheries in the form of CD ROM database, software, etc.

Services Rendered to Scientific Communities

Besides training scientists/research workers through periodical workshops, regular services were offered by way of extending the following facilities. Expert consultation in statistical design, analysis and interpretation of aquaculture data, Presentation package development, Training on computer application in Aquaculture, Bibliographic search services (Online & offline through CD-ROM), E-mail, Internet browsing, CDROM search, Scanning & Printing,

Acknowledgements

Authors are thankful to Dr. T. Madhan Mohan, Adviser, Bioinformatics Division, Department of Biotechnology, Ministry of Science and Technology, Govt. of India, New Delhi for funding the project.

References

Anon, 2005. Aquaculture Technologies for Farmers. Indian Council of Agricultural Research, Krishi Bhavan, New Delhi-110001. PP.104.

Anon, 2006. Final Report (RPF-III). April 1996 to March, 2007. Social Science Research in Aquaculture, CIFA, Kausalyaganga. pp. 132.

Murthy, Lakshmi, 2006. Networking Professionals through Discussion Forums on the Virtual Platform. In: Bioinformatics and Statistics in Fisheries Research. Vol.2 (Eds. A. K. Roy, A. Antony and N. Sarangi), CIFA, Kausalyaganga, Bhubaneswar. pp.185-192.

Roy, A.K., K.N. Sahoo, K.P. Sahoo and G.S. Saha (2001). A Socio-economic study on aquaculturists of Kolleru Lake, Andhra Pradesh. *In*: Eco-friendly management of resources for doubling fish production- strategies for 21st century. (Eds. M. Sinha, D. Kumar and P. K. Kathia). Inland Fisheries Society of India, Barrackpore pp. 176-179.

Roy, A. K., Saha, P. K., Datta, A. K. and Satapathy, P. K. 2000. Regression model for *Catla catla* reared in sewage-fed ponds. In: *Waste Recycling and Resource Management in the Developing World* (Eds. B. B. Jana, R. D. Banerjee, B. Guterstam and J. Heeb). University of Kalyani, India and International Ecological Engineering Society, Switzerland, pp.65-69.

Gupta, S. D., Roy, A. K. Rath, S. C., Satapathy, P. K., 1999. Database in induced breeding experiment on an Indian major carp, *Labeo rohita* (Ham.). *In*: Bioinformatics and Statistics in Fisheries Research (Eds. A.K.Roy, A.Antony and M.Rout), Central Institute of Freshwater Aquaculture, Bhubaneswar, and pp: 117-118.

12

Concepts in Development of Bioinformatics Towards Systems Biology

A. N. Misra and M. Misra

The advancements in the field of informatics, biology and allied fields of technology led to the emergence of an inter-disciplinary field known as 'Bioinformatics'. This field is an expanding and challenging tool for the collection, storage, organization, integration, analysis and simulation of biological data for utilization in biotechnology.

Collection of Data for Bioinformatics

An epoch making revolution in Genomics in recent years in deciphering plant, animal and microbial genome sequences, Proteomics, Statistics, Instrumental techniques, Structural and Computational Biology have given input to fragmented pieces-and-bits of information on biological materials. Each data is a discrete piece of important but defragmented information regarding a molecule (protein, RNA, DNA, sugars or lipids etc) or a reaction or at best a pathway, supposed to be occurring in a living system independently at a given time and space. However, as in a living system each molecule or a process exists in unison with other entities of its kind, these fragmented data are to be stored, organized, and then integrated and analysed to form an integrated part of a whole chain of events – now called a 'SYSTEM'. Storage and organisation of data Explosion of knowledge in the field of biology and generation of new informations (data) needs well storage, organization and a faster distribution among the scientific workers. The advent of super computers, high speed computing and development of algorithms for data formatting, assists in the storage of data at the site of origin and distributed through the hyperlinks in world wide web (www).

Genomics and Proteomics

The vast field of Bioinformatics is beyond the scope of this article. However to elucidate the concepts that led to the the development of Bioinformatics is given in brief. Both genomics and proteomics deals with comparative, functional and structural aspects. Comparative genomics or proteomics entails the comparison of new and known sequences with the available sequence database available in genome or proteome databases in NCBI, PubMed, SWISS PROT, EMBL etc. There are hundreds of such databases in public domain and are now available commercially for specific uses. Functional genomics, proteomics and the interface of these two – the *TRANSCRIPTOMICS* uses the expression patterns and profiles. It started with individual gene or RNA or protein expression using blotting (Southern, Northern or Western) techniques culminating in the use of *chips* for global expression patterns. The structural studies encompass the 3-D structure elucidation using X-ray diffraction, CD and other spectroscopic studies, and computation analysis of molecular stability.

Integration and Analysis

Data collected in a piece meal basis from among all schools of thought and approach is collected, stored and then integrated into a organized into well defined groups for further.analysis. Now it is defined as a 'data base'. The integration of data needs some logical inputs and computational approach. Analysis of data is done when such logical distribution of data is possible to access different molecules or function and structure as per the user defined manner. In a integrated data base analysis can be done in a 'bottom up' or 'top down' approach. The bottom up approach or disintegration approach is an analytical tool for understanding individual or few data at a point of time. In contrast, the top down approach is an integrated approach of analysing a group of molecules and their interactions or a network of events which occurs *in vivo* in a cell or an organism. Such analysis led to the evolution of the Systems approach in Genomics, transcriptomics, and proteomics, with new field emerging as metabolomics and glycomics.

Simulation of Data and Systems Biology

System biology is an integrative approach in the study of a system as a whole. With this approach the data collected in a fragmented manner, which is a reductionist approach in collecting fragments of the system, can be simulated through bioinformatics for constructing a structural and functional model in a cellular or whole organism system. These interactive - systems approach can help in the development of commercial biodesigns and targets *in silico*.

Emerging Areas of Bioinformatics for Systems Biology

It will not be contrary to say that Bioinformatics is still at infancy. The integration of bioinformatics with systems biology in the era of biotechnology is progressing fdast for commercial exploitation of the available biological data. New areas which is emerging with these Systems approach to Bioinformatics are Interactomics (protein protein, protein-sugar, protein-nucleic acid etc.), Fluxomics or Phenomics (kinetic changes), and Biomics (Biome analysis). These are few of the vast array of emerging trends and many more to come in near future with heavy demand in the Biotech, Pharmaceutical and Genomics industry.

References

Bains, W. 2004. Biotechnology from A to Z. 3 rd Edn., Oxford Univ. Press, New York.

Kumar, S. 2006. Basics in Bioinformatics. *In* Bioinforamtics & Statistics in Fisheries Research. Eds. Roy AK, Antony A and Sarangi N, CIFA, Bhubaneswar, pp.10-21.

Madan Mohan, T. 2006. Growth of Biotchnology information network (BTIS net) in India. *In:* Bioinforamtics & Statistics in Fisheries Research. Eds. Roy AK, Antony A and Sarangi N, CIFA, Bhubaneswar, pp.3-9.

Mehrotra P, Sarin K. and Srivastava 2005. The New Handbook of Bioinformatics. 1st Edn., Vikas Publ. House, New Delhi.

13

Nanotechnology in Bioinformatics

L. Kanungo

Nanotechnology is a field of applied science focused on the design, synthesis, characterization and application of materials and devices on the nanoscale. It is a sub classification of technology in colloidal science, biology, physics, chemistry and other scientific fields and involves the study of phenomena and manipulation of material at the nanoscale. This is involved in the creation of useful materials, devices, and systems used to manipulate matter at an incredibly small scale; between 1 and 100 nanometers. With nanotechnology, a large set of materials with distinct properties (optical, electrical, or magnetic) can be fabricated. Nanotechnologically improved products rely on a change in the physical properties when the feature sizes are shrunk.

Achievements through Nanotechnology

Nanotechnology involves the creation and use of materials and devices at the level of molecules and atoms. Research in nanotechnology began with applications outside of medicine and is based on discoveries in physics and chemistry. This is because it is essential to understand the physical and chemical properties of molecules or complexes of molecules in order to control them. The same holds true for the molecules and structures inside living tissues. Researchers have developed powerful tools to extensively categorize the parts of cells in vivid detail, and we know a great deal about how these intracellular structures operate.

Nanotechnology in medicine

A nanometer is one-billionth of a meter, too small to be seen with a conventional lab microscope. It is at this size scale, about 100 nanometers or less that biological molecules and structures inside living cells operate. From nanotechnology it is only one step to nanomedicine, which may be defined as "the monitoring, repair, construction and control of human biological systems at the nanolevel using nanodevices and nanostructures". Nanomedicine, an offshoot of nanotechnology, refers to highly specific medical intervention at the molecular scale for curing disease or repairing damaged tissues, such as bone, muscle, or nerve. Proper diagnosis and efficient delivery of drugs are very necessary in the medical fields where nanosize materials have practical implementations. Such nanotools still await construction and at present, they are in the progress. Nevertheless, they might be very helpful and become a really diagnostic tool in the near future. The use of nanodevices such as the artificial RBCs-respirocytes will definitely solve the problem of thalassemia patients, which has become a curse to these people.

Nanobiosensors

Nanobiosensor is a probe that integrates a biological component, such as a whole bacterium or a biological product (e.g., an enzyme or antibody) with an electronic component to yield a measurable signal. The nanoprobes are fabricated with optical fibers pulled down to tips with distal ends having sizes of approximately 30-50 nm. The nanoscale size of these sensors allows for measurements at single cell level. The fiber optic nanoprobes are covalently bound with antibodies that are selective to target analyte molecules. Excitation light is launched into the fiber and the resulting transient field at the tip of the fiber is used to excite target molecules bound to the antibody molecules. The florescence emission from the analyte molecule is then connected to a microscope. As aging diseases such as diabetes and Alzheimer's, the chemical warfare agents cause changes in metal ion concentrations in the body, these changes can be detected and measured by nanobiosensors, and may help in availing information about changes in disease states and exposure to toxins. A new type of biosensor has been developed, that can detect the macromolecules in nanomolar concentration. Application of fiber optic based nanobiosensors for in vivo measurement of carcinogen benzo (a) pyronine can potentially provide both quantification and identification of protein analyte at single molecule level.

Nanotubes

Carbon nanotubes (CNT) exhibit unique electronic and extraordinary mechanical properties. They are only 1 nm in diameter in the form of films and aligned bundles. The tip of the nanotubes can be functionalised with appropriate probe molecules for diagnostics. A prototype catheter can be developed which would permit detection of specific oligonucleotide sequences that serve as molecular signatures of cancer cells. A new class of antibacterial peptides is being developed. Nanotubes are formed by self-assembly of cyclic peptides composed of alternating D- and L-amino acids. With appropriate design, the nanotubes insert themselves into bacterial, but not mammalian, cell membranes. Pores are created, resulting in bacterial cell death.

Nanorobots

The other important application of nanotechnology is the design of nanorobots, which can enter inside the body and repair the tissue. Thus, the use of nanotools and nanomachines will be possible not only for laboratory experiment but also for the health point of view in the near future.

Nanotechnology and Cancer Therapy

In the fight against cancer, nanotechnology introduces unique approaches to diagnosis and treatment that could not even be imagined with conventional technology. New tools engineered at sizes much smaller than a human cell will enable researchers and clinicians to detect cancer earlier, treat it with much greater precision and fewer side effects, and possibly stop the disease long before it can do any damage. A nanoparticle can be used to light up a tumor in an MRI, destroy cancer cells by converting magnetic fields into heat, and allow the physician to visually track the progress of treatment. Nanotechnology offers the unprecedented and paradigm-changing opportunity to study and interact with normal and cancer cells in real time, at the molecular and cellular scales, and during the earliest stages of the cancer process. Through the concerted development of nanoscale devices or devices with nanoscale materials and components, the NCI Alliance for Nanotechnology in Cancer will facilitate their integration within the existing cancer research infrastructure. The Alliance will bring enabling technologies for

1. Imaging agents and diagnostics that will allow clinicians to detect cancer in its earliest stages.
2. Systems that will provide real-time assessments of therapeutic and surgical efficacy for accelerating clinical translation.
3. Multifunctional, targeted devices capable of bypassing biological barriers to deliver multiple therapeutic agents directly to cancer cells and those tissues in the microenvironment that play a critical role in the growth and metastasis of cancer.
4. Agents that can monitor predictive molecular changes and prevent precancerous cells from becoming malignant.
5. Novel methods to manage the symptoms of cancer that adversely impact quality of life.
6. Research tools that will enable rapid identification of new targets for clinical development and predict drug resistance.

Because of the small size of nanoscale devices, they can readily interact with biomolecules on both the surface of cells and inside of cells. By gaining access to so many areas of the body, they have the potential to detect disease and deliver treatment in ways unimagined before now. And since biological processes, including events that lead to cancer, occur at the nanoscale at and inside cells, nanotechnology offers a wealth of tools that are providing cancer researchers with new and innovative ways to diagnose and treat cancer. It can provide rapid and sensitive detection of cancer-related molecules by enabling scientists to detect molecular changes even when they occur only in a small percentage of cells. Nanoscale devices have the potential to radically change cancer therapy for the better and to dramatically increase the number of highly effective therapeutic agents. Nanoscale constructs can serve as customizable, targeted drug delivery vehicles capable of ferrying large doses of chemotherapeutic agents or therapeutic genes into malignant cells while sparing healthy cells, greatly reducing or eliminating the often unpalatable side effects that accompany many current cancer therapies. Examples of nanotechnology in cancer research today include the following:

- Nanoscale cantilevers and nanowire sensors can detect biomarkers of cancer from a single cell, which heretofore was unimaginable.
- Nanoparticles can aid in imaging malignant lesions, so surgeons know where the cancer is, and how to remove it.
- Nanoshells can kill tumor cells selectively, so patients don't suffer terrible side effects from healthy cells being destroyed.
- Dendrimers can sequester drugs to reduce systemic side effects, deliver multiple drugs to maximize therapeutic impact, and rapidly discern effectiveness of a drug.

Conclusion

Concerns have been voiced that nanotechnology, specifically biomedical nanotechnology, may be advancing too fast, and regulatory agencies are not able to adequately address potential safety concerns. The NCI recognizes the importance of advancing our knowledge and understanding of the technology, while also ensuring that the technology is safe and effective. The success of efforts in the field is contingent upon scientific excellence in research and development that is both ethical and safe for the body and the environment. The NCI is systematically addressing these issues within the purview of its biomedical expertise and its collaborations, primarily through the newly developed Nanotechnology Characterization Laboratory. The NCI has partnered with the National Institute of Standards and Technology (NIST) to help develop criteria to define all the key physical and biological characteristics of nanodevices intended for use in cancer-related nanotechnology research. Furthermore, the NCI is expanding its interaction with the FDA to facilitate smoother transition from bench-to-bedside by ensuring that researchers have sufficient and appropriate data to guide the development of safe and efficient nanodevices.

References

Cameron, N.M. 2006. Nanotechnology and the Human Future: Policy, Ethics, and Risk. *Ann N Y Acad Sci.*, 1093:280-300. PMID: 17312264.

Buxton, D.B. 2007. Nanotechnology in the diagnosis and management of heart, lung and blood diseases. *Expert Rev Mol Diagn.*, 7(2):149-60. PMID: 17331063.

Gordon, A.T., Lutz, G.E., Boninger, M.L., Cooper, R.A. 2007. Introduction to Nanotechnology: Potential Applications in Physical Medicine and Rehabilitation. *Am J Phys Med Rehabil.* 2007 Mar; 86(3): 225-241.PMID: 17314708.

Ehdaie B. 2007. Application of Nanotechnology in Cancer Research: Review of Progress in the National Cancer Institute's Alliance for Nanotechnology. *Int J Biol Sci.* 25;3(2):108-10.

Wilson RF. 2006. Nanotechnology : the challenges of regulating known unknowns. *Nanomedicine.* 2(4):318. PMID: 17292298.

Koo Om, Rubinstein I, Onyuksel H. 2005. Role of nanotechnology in targeted drug delivery and imaging : a concise review. *Nanomedicine*, 1(3):193-212. PMID: 17292079.

Roco MC. 2006. Progress in governance of converging technologies integrated from the nanoscale. *Ann N Y Acad Sci.* 1093:1-23. PMID: 17312249.

14

Use of Biosensors in Bioinformatics

R. R. Patra

The advances in science and technology are deeply intertwined. Researchers can use the semiconductor manufacturing techniques that underlie miniaturization to build radios and exceptionally small mechanical structures that sense fields and forces in the physical world. These inexpensive, low-power communication devices can be deployed throughout a physical space, providing dense sensing close to physical phenomena, processing and communicating this information, and coordinating actions with other nodes. These multifunction devices are battery operated and are connected wirelessly with sensing, processing and communicating capabilities and are known as the wireless sensors. If we use these sensors with biological issues or Biotechnology, then we can find the new concepts biosensors. If we really implement these concepts in real life we can get rid of many diseases like heart attack, blood sugar level, gene reaction and other changes that happens in our body much before its shows to outside.

A wireless sensor network (WSN) is composed of a large number of sensor nodes, which are densely deployed in an ad hoc fashion either inside the phenomenon or very close to it. The sensor nodes are usually scattered in a wireless sensor network randomly. Each of these scattered sensor nodes has the capabilities to collect data and route data back to the *sink*. Data are routed back to the sink by a multihop infrastructure less architecture through the sink. A sensor network design is influenced by many factors, which include fault tolerance; scalability; production costs; operating environment; sensor network topology; hardware constraints; transmission media; and power consumption. However, none of these studies has a full-integrated view of all factors that are driving the design of sensor networks and sensor nodes. These factors are important because they serve as a guideline to design a protocol or an algorithm for sensor networks. It is observed that some nodes in WSN may fail or be blocked due to lack of power, have physical damage or environmental interference. The failure of sensor nodes should not affect the overall task of the sensor network. This is the reliability or fault tolerance issue. Fault tolerance is the ability to sustain sensor network functionalities without any interruption due to sensor node failures.

A sensor node's coverage can be described as every point in space that lies within its sensing range. Aggregation of sensor-coverage from each sensor node provides an estimate of the total sensor coverage. Sensors are generally deployed randomly and there is high probability that some region of the network is not covered, whereas some other regions have redundant sensors. It is very essential for the wireless sensor network to be well covered and to maintain the connectivity in order to transmit the important sensed data time to time. In such a multi-hop communication model, the size of the data has a direct effect on how efficient the network may be in conserving energy. The main steps for energy-efficient coverage estimation include reducing the amount of data required to describe individual sensor-coverage, modeling communication schemes that efficiently use properties of data, and distributing the task load among the nodes.

In a post deployment scenario, the coverage and connectivity of the nodes may be affected due to several factors such as reach ability i.e. due to jamming, noise, moving obstacles, etc. or death of the nodes due to energy scarcity, accidental death due to software or hardware malfunctioning. Sensor nodes may be statically deployed. However, device failure is a regular or common event due to energy depletion or destruction. It is also possible to have sensor networks with highly mobile nodes. Besides, sensor nodes and the network experience varying task dynamics, and they may be a target for deliberate jamming. Therefore, sensor network topologies are prone to frequent changes after deployment. In most of the applications, these nodes are deployed in the ocean surface, under the water, in the forest or in inaccessible terrains, in human body, on any plant or any living or moving elements and very difficult to maintain the network connectivity and coverage manually. Biosensors are future concepts by which we can see a better world in health and prosperity.

15

Applications of Statistics in Bioinformatics

A.K. Roy and S.R. Martha

Introduction

After the advent of Post Genomic era, Biological data is accumulating at an electrifying pace every day. Storage and analysis of these data serve as a major source for scientific research to understand the phenomena of origin, evolution and existence of various forms of life. The management of these data requires different statistical techniques used for retrieval, storage, analysis and extraction of precise information from small as well as large biological datasets. Application of statistical tools in Bioinformatics especially in fisheries data has been reported (Roy and Martha, 2008).

All raw data need to undergo the following few stages for generation of useful information and knowledge discovery. The process of knowledge discovery via data mining can be divided into four basic activities; *selection, pre-processing, data mining*, and *interpretation*. Selection involves creating the target data set, i.e. the data set to about undergo analysis. As discussed previously, modern datasets may be both large and complex. The larger the amount of available data, the greater is the likelihood that an identifiable trend or pattern may be recognized and statistically validated. If the dataset is relatively complex, it is often considered impractical to attempt to subject the complete dataset for analysis. Pre-processing involves preparing the dataset for analysis by the data mining software. This may involve resolving undesirable data characteristics such as missing data (non-complete fields), irrelevant fields, non-variant fields, skewed fields, and outlying data points. The pre-processing activities may result in the generation of a number of (potentially overlapping) subsets of the original target dataset. Preprocessing is process of filtering of data to avoid redundancy (repetition) and missing of value in between. Thereafter cleaned data is subjected to analysis by the data mining software in an attempt to identify hidden trends or patterns, or to test specific hypotheses. This involves the application of different statistical analysis methods. Finally interpretation and of the results produced is carried out. This may well involve returning to previous stages to carry out additional activities in order to provide further information if necessary.

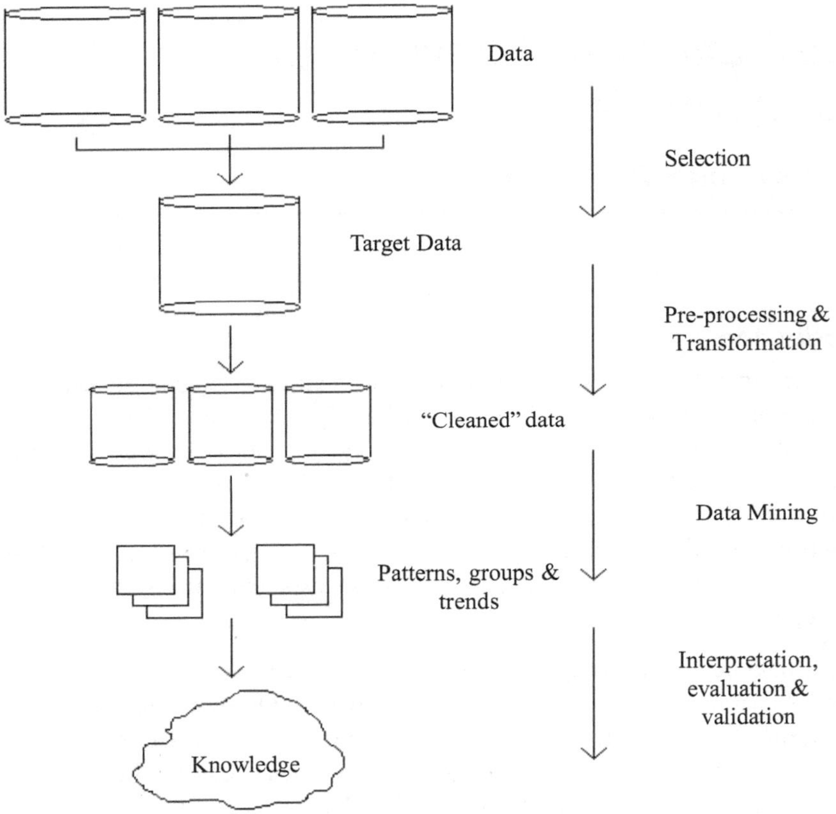

Fig. 1 : Overview of Knowledge Discovery

Statistical Data Mining

Data Mining is all about automating the process of searching for patterns in the data. It is the process of working on problems which are a composition of four underlying subproblems: classification, clustering, pattern search and outlier detection. Different Methods of Statistical Data mining are listed below that is hosted by Andrew Moore:

- Decision Trees
- Information Gain
- Probability for Data Miners

- Probability Density Functions
- Gaussians
- Maximum Likelihood Estimation
- Gaussian Bayes Classifiers
- Cross-Validation
- Neural Networks
- Instance-based learning (aka Case-based or Memory-based or non-parametric)
- Eight Regression Algorithms
- Predicting Real-valued Outputs: An introduction to regression
- Bayesian Networks
- Inference in Bayesian Networks (by Scott Davies and Andrew Moore)
- Learning Bayesian Networks
- A Short Intro to Naive Bayesian Classifiers
- Short Overview of Bayes Nets
- Gaussian Mixture Models
- K-means and Hierarchical Clustering
- Hidden Markov Models
- VC dimension
- Support Vector Machines
- PAC Learning
- Markov Decision Processes
- Reinforcement Learning
- Elementary probability and Naive Bayes classifiers
- Spatial Surveillance
- Time Series Methods
- Game Tree Search Algorithms, including Alpha-Beta Search
- Zero-Sum Game Theory
- Non-zero-sum Game Theory

- Introductory overview of time-series-based anomaly detection algorithms
- AI Class introduction
- Search Algorithms
- A-star Heuristic Search
- Constraint Satisfaction Algorithms, with applications in Computer Vision and Scheduling
- Robot Motion Planning
- Hill Climbing, Simulated Annealing and Genetic Algorithms

Out of these methods few are Decision Tree, Gaussians, Maximum Likelihood Estimation, Neural Networks, Bayesian Network, Hidden Markov Models, Time Series Methods, Hierarchical clustering, K-means, self-organizing maps (SOMs), Correspondence Analysis and Principal Component Analysis (PCA) commonly used in bioinformatics which are described in brief for better understanding (Moore, 2008).

Decision Tree

A decision tree (or tree diagram) is a decision support tool that uses a graph or model of decisions and their possible consequences, including chance event outcomes, resource costs, and utility. A decision tree is used to identify the strategy most likely to reach a goal. These trees serve as a descriptive means for calculating conditional probabilities. They also help you to form a balanced picture of the risks and rewards associated with each possible course of action. A Decision Tree is a tree-structured plan of a set of attributes to test in order to predict the output.

Gaussian

Gaussians, both the friendly univariate kind, and the slightly-reticent-but-nice-when-you-get-to-know-them multivariate kind are extremely useful in many parts of statistical data mining, including many data mining models in which the underlying data assumption is highly non-Gaussian. Gaussian method is also known as Normal Distribution of Bell-shaped curve.

General Gaussian — Also known as the normal distribution or Bell-shaped curve

$$p(x) = \frac{1}{\sqrt{2\pi\sigma}} \exp\left(\frac{(x-\mu)^2}{2\sigma^2}\right)$$

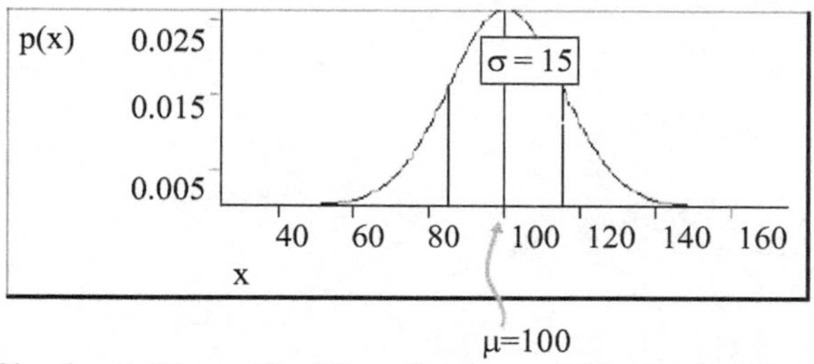

$E[X] = \mu$

$Var[X] = \sigma^2$

Shorthand: We say $X \sim N(\mu, \sigma^2)$ to mean "X is distributed as a Gaussian with parameters m and σ^2"

In the above figure, $X \sim N(100, 15^2)$

Fig. 2: General Gaussian Method

Maximum Likelihood Estimation

Maximum likelihood estimation (MLE) is a popular statistical method used to calculate the best way of fitting a mathematical model to some data. Modeling real world data by estimating maximum likelihood offers a way of tuning the free parameters of the model to provide an optimum fit.

A General MLE strategy

Suppose $\theta = (\theta_1, \theta_2, ..., \theta_n)T$ is a vector of parameters.

Task: Find MLE θ assuming know form for p(Data | θ, stuff)

1. Write LL = log P(Data | θ, Stuff)
2. Work out $\partial LL/\partial\theta$ using high-school calculus
3. Solve the set of simultaneous equations

If you can't solve them, what should you do?

$$\frac{\partial LL}{\partial \theta_1} = 0$$

$$\frac{\partial LL}{\partial \theta_2} = 0$$

4. Check that you're a maximum

Fig. 3: An example of Maximum Likelihood Estimation

Neural Networks

A Neural Network (NN) is an information processing paradigm that is inspired by the way biological nervous systems, such as the brain, process information. The key element of this paradigm is the novel structure of the information processing system. It is composed of a large number of highly interconnected processing elements (neurons) working in union to solve specific problems. An

Artificial Neural Network (ANN) is configured for a specific application, such as pattern recognition or data classification, through a learning process. Learning in biological systems involves adjustments to the synaptic connections that exist between the neurons. This is true of ANNs as well.

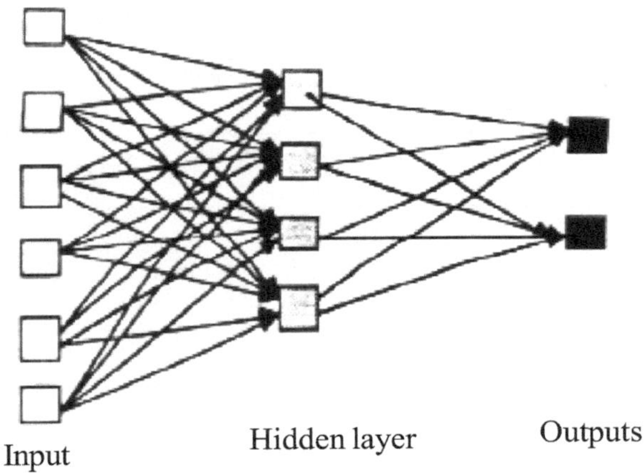

Fig. 4: An example of a simple feed forward neural network

Bayesian Network

A Bayesian network is a graphical model that encodes probabilistic relationships among variables of interest. When used in conjunction with statistical techniques, the graphical model has several advantages for data analysis. The model encodes dependencies among all variables, it readily handles situations where some data entries are missing. A Bayesian network can be used to learn causal relationships, and hence can be used to gain understanding about a problem domain and to predict the consequences of intervention. The model has both a causal and probabilistic semantics, it is an ideal representation for combining prior knowledge (which often comes in causal form) and data. Bayesian statistical methods in conjunction with bayesian networks offer an efficient and principled approach for avoiding the overfitting of data.

K-Means Clustering

The -means algorithms randomly select k data points as initial means. K clusters are formed by assigning each data point to its closest cluster mean. The algorithm uses the Eucledian distance. Virtual means for each cluster are calculated by using all data points contained in a cluster. In any K-means clustering algorithm defines, the user defines the number of clusters depending upon which a K-cluster center locations are formed after which each data point finds out the center it is closest to. Thereafter each center find the centroid of the points by its own.

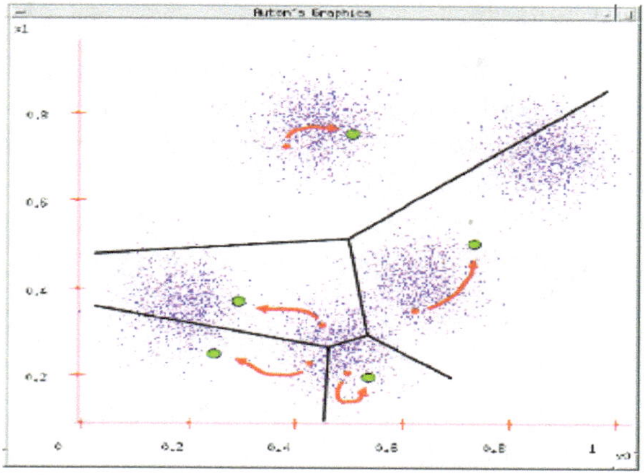

Fig. 5: An example of K-means clustering showing 5 clusters (*Source*: Andrew)

Hidden Markov Models

A hidden Markov model (HMM) is a statistical model in which the system being modeled is assumed to be a Markov process with unknown parameters, and the challenge is to determine the hidden parameters from the observable parameters. The extracted model parameters can then be used to perform further analysis, for example for pattern recognition applications. A HMM can be considered as the simplest dynamic Bayesian network

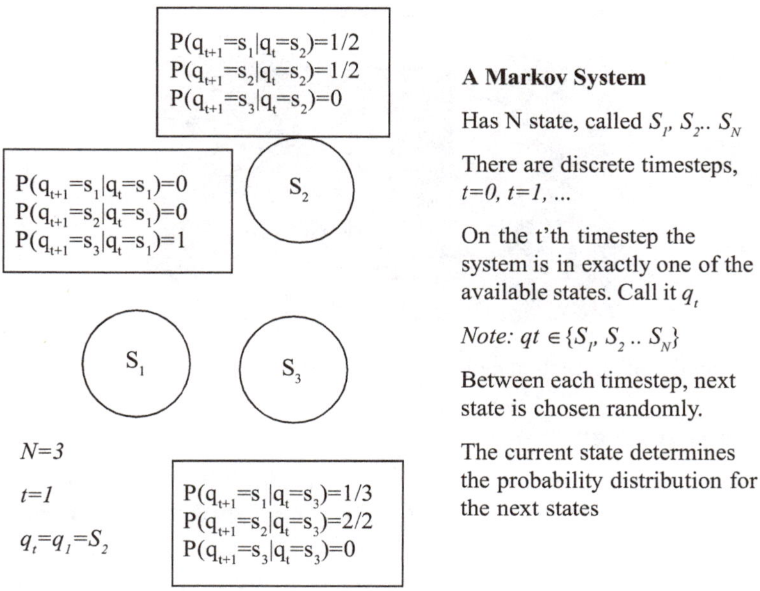

Fig. 6 : An example of Hidden Markov Model

Time Series Methods

The term "univariate time series" refers to a time series that consists of single (scalar) observations recorded sequentially over equal time increments. Although a univariate time series data set is usually given as a single column of numbers, time is in fact an implicit variable in the time series. If the data are equi-spaced, the time variable, or index, does not need to be explicitly given. The time variable may sometimes be explicitly used for plotting the series. However, it is not used in the time series model itself.

Hierarchical Clustering

The hierarchical clustering can be represented as a tree, or a dendrogram. Branch lengths represent the degree of similarity between the genes. Hierarchical clustering consists of two separate phases. The hierarchical clustering is performed as follows; Each gene is assigned to a cluster of its own. The closest pair of clusters is found and merged into a single cluster. The distances (similarities) between the new clusters are computed and each of the old clusters using either single, average or complete linkage method. Above steps are repeated until all genes are clustered.

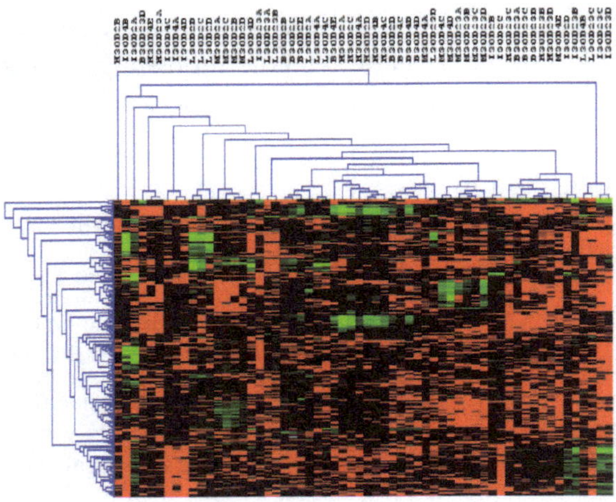

Fig. 7 : An example of Hierachial clustering of C. carpio obtained using MeV Software (*Source*: Roy *et al.*, 2007)

Self-Organizing Map

Kohonen's self-organizing map (SOM) is a neural net that uses unsupervised learning. SOM tries to learn to map similar input vectors (gene expression profiles) to similar regions of the output array of nodes. The method maps the multidimensional distances of the feature space to two-dimensional distances in the output map.

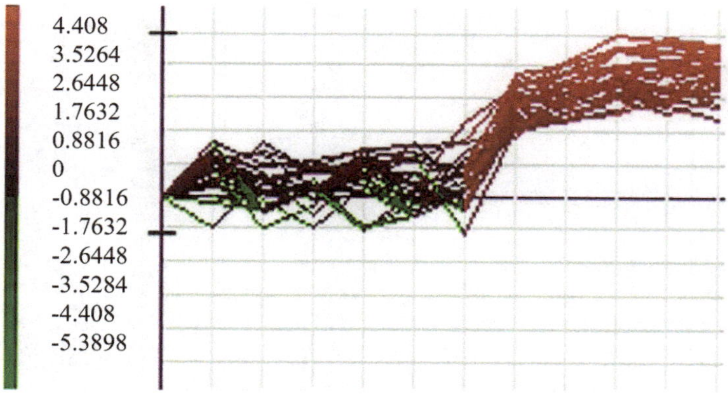

Fig. 8 : An example of self-organizing map (*Source*: Hovatta *et al.*)

Principal Component Analysis

Objectives of principal component analysis (PCA) are to discover or to reduce the dimensionality of the data set and to identify new meaningful underlying variables. PCA transforms a number of (possibly) correlated variables into a (smaller) number of uncorrelated variables called principal components. The basic idea in PCA is to find the components that explain the maximum amount of variance possible by n linearly transformed components. The first principal component accounts for as much of the variability in the data as possible, and each succeeding component accounts for as much of the remaining variability as possible.

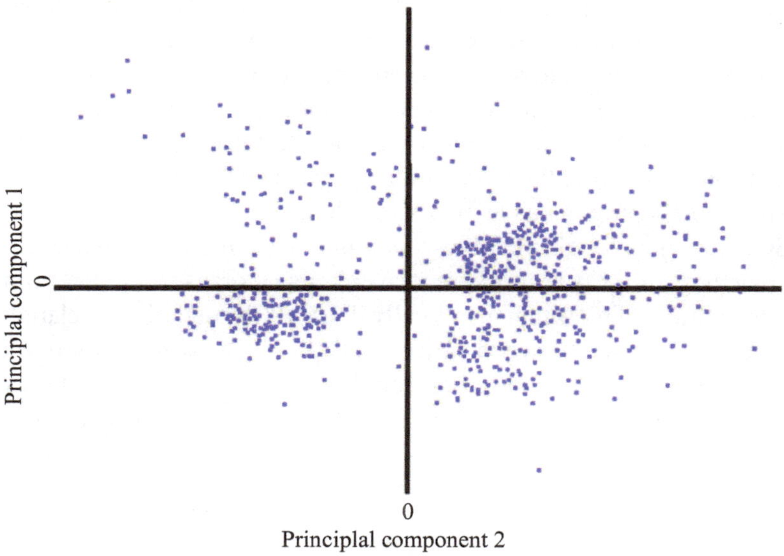

Fig. 9 : An example of principal component analysis. The two most significant principal components have been selected as the axes of the plot (*Source*: Hovatta *et al.*)

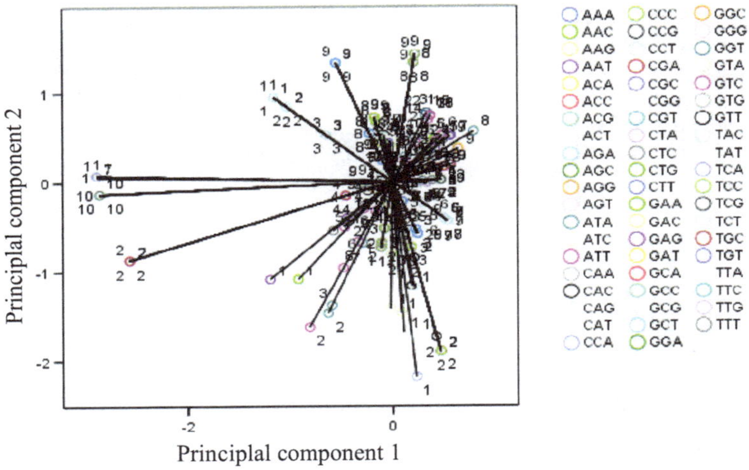

Fig. 10 : An example of Principal Component Analysis generated by SPSS.

Correspondence Analysis

Correspondence Analysis is a multivariate exploratory graphical technique used to describe the relationship between the row and column variables of a contingency table. It gives output in form of correspondence table, table of dimensionality, profiles and distances, row and column scores, eigen values, mass, centroid, singular values, confidence statistics, plots of symmetrical normalization. A correspondence table is any two-way table whose cells contain some measurement of correspondence between the rows and the columns. The measure of correspondence can be any indication of the similarity, affinity, confusion, association, or interaction between the row and column variables. It generates results in the form of tables and graphs. It is used in case of genomic sequence analysis to find the association between genes and codons. A matrix is constructed keeping genes as row variable and codons as column variable and RSCU values as weight variables that is uploaded into the statistical package SPSS to carry out Correspondence Analysis. Correspondence Analysis generates a variety of plots that graphically illustrate the underlying relationships between categories and between variables. Biplot is the scatter plot of the row and column scores for the two-dimensional solution. A number of studies have been made on application of CoA in Fish Genome Studies (Roy *et al.*, 2007; Roy & Martha, 2008).

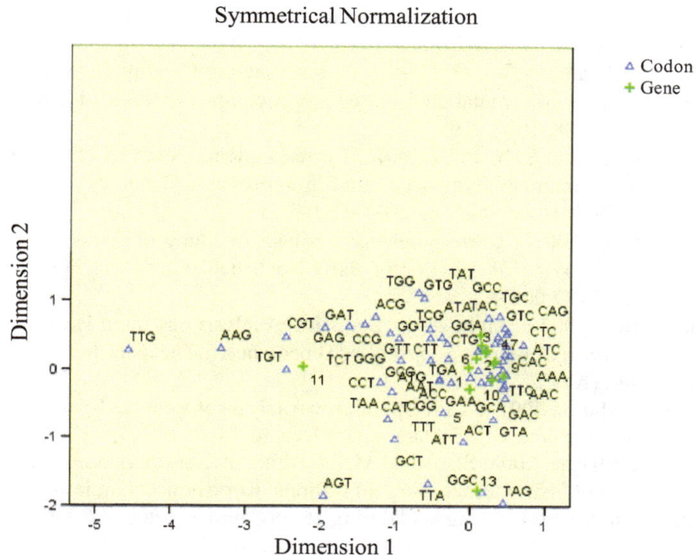

Fig.11 : Biplot between genes and codons of C. carpio generated by Correspondence Analysis of SPSS (*Source*: Roy *et al.*, 2008)

Conclusion

Demonstration of statistical methods with illustrated examples will help future researchers in carrying out complex statistical analysis using statistical packages and bioinformtics tools and techniques to visualize fact through data mining and exploratory data analysis. For further knowledge the researchers may refer to the following bibliographics.

References

Greenacre, M.J. (1984). Theory and Applications of Correspondence Analysis. Academic Press, London.

Holm, L. (1986). Codon usage and gene expression. *Nucleic Acid Res.*, 14: 3075- 3087.

Larkin, Patrick; Villeneuve, Daniel L; Knoebl, Iris; Miracle, Ann L; Carter, Barbara J; Liu, Li; Denslow, Nancy D; Ankley, Gerald T. (2007) Development and Validation of a 2,000-Gene Microarray for the Fathead Minnow (Pimephales Promelas). *Environ. Toxicol. Chem.* 26(7):1497-1506.

McLean, TI; Crawford, DL. (2006). Towards a Genetic Understanding of Karenia brevis: Using Microarrays to Discover Gene Expression Patterns as They Relate to K. brevis Ecology EOS, Transactions, American Geophysical Union [EOS Trans. Am. Geophys. Union]. 87(36): suppl., [np]. suppl.

Mclnerney, J.O. (1998). GCUA: General Codon Usage Analysis. Bioinformatics, 14:372-373. Medigue, C., T. Rouxel, P. Vigier, A. Henaut and Danchin, A. (1991). Evidence for horizontal gene transfer in Escherichia coli speciation. *Journal of Molecular Biology*, 222: 851-856.

Morrison, D.A., Ellis, J., and Johnson, A.M. (1994) An empirical comparison of distance matrix techniques for estimating codon usage divergence. *Journal of Molecular Evolution* 39:533-536.

Osawa, S. and Jukes, T. H. (1989). Codon reassignment (codon capture) in evolution. *Journal of Molecular Evolution*, 29: 271-278.

Romero, H., Zavala, A. and Musto, H. (2000). Codon usage in Chlamydia trachomatis is the result of strand-specific mutational biases and a complex pattern of selective forces. *Nucleic Acid Res.*, 28: 1084-2090.

Roy, A. K, S. R. Martha and S. N. Rath. 2007a. Correspondence Analysis of Codon Usage in Labeo rohita, International Symposium on Chromosome to Genomes (c2g), held on 3-5 July 2007 (CCMB, Hyderabad, India). 9-10 (Abs.).

Roy, A. K. and Martha. 2007b. Correspondence Analysis in Study of Genomic Sequence of Labeo rohita, 8th Asian Fisheries Forum (8aff), Kochi held during 20-23 November 2007. 317-318 (Abs.GMO 008).

Roy, A. K. and Martha. 2007 c. Multivariate Statistical Analysis of Codon Usage values of L. rohita. International Conference on Applied Bioengineering held at Chennai during 5th-7th December 2007(Abs.).

Roy, A. K. and S. R. Martha 2007 d. Correspondence Analysis of Relative Synonymous Codon Usage values of L.rohita. SPSS Analyst. Oct-Dec:30-35.

Roy, A. K. and S. R. Martha. 2008. Statistical Methods for Analysis of Genomic Sequence and Microarray data of Fishes. Proceedings of National workshop cum training on Molecular Modelling and Protein Docking held during 25-28 February 2008 at CARI. 3-17.

Thioulouse, J., Chessel, D. and Dolédec, S. (1997). ADE-4: A multivariate analysis and graphical display software. Stat. Comput., 7: 75–83.

Weidinger, G; Thorpe, CJ; Wuennenberg-Stapleton, K; Ngai, J; Moon, RT*.(2005). The Sp1-Related Transcription Factors sp5 and sp5-like Act Downstream of Wnt/ beta -Catenin Signaling in Mesoderm and Neuroectoderm Patterning Current Biology [Curr. Biol.]., 15(6): 489-500.

http://www.autonlab.org/tutorials/

http://www.niedermayer.ca/papers/bayesian/ www.cs.brown.edu/research/ai/dynamics/tutorial/Documents/HiddenMarkovModels.html www.doc.ic.ac.uk/~nd/surprise_96/journal/vol4/cs11/report.html

16

Algorithms in Sequence Alignment

R. Dash

Biological sequences like the nucleotide and amino acid sequences can be analyzed by means of the sequence alignment. The objective behind this is to obtain knowledge regarding the similarity between the query and the target sequence to decipher the structure, function, homology of the query sequence.

Sequence Alignment can be categorized into 2 types

1. Local alignment
2. Global alignment

Local alignment

An alignment that searches for segments of the two sequences that match well. There is no attempt to force entire sequences into an alignment, just those parts that appear to have good similarity, according to some criterion. Example:

```
aaagcggaagtcacag
......|||||....
aaggctgaagt-atag
```

Global alignment

An alignment that assumes that the two proteins are basically similar over the entire length of one another. The alignment attempts to match them to each other from end to end, even though parts of the alignment are not very convincing. Example:

```
aaagcggaagtcacag
||.||.||||| |.| |
aaggctgaagt-atag
```

Alignment Methods

- Visual
- Brute Force
- Dynamic Programming
- Word-Based (k tuple)
- Visual Alignments - Dot Plots

Dot plots of sequence similarity are created using a matrix where

- rows in the matrix correspond to the characters in the first sequence,
- columns in the matrix correspond to the characters in the second sequence.

Steps

- Loop through each row.
- For the current row, take the character in that row and compare it to the character in each column. If they are equal, place a dot in the matrix.
- Continue until all nodes in the matrix have been considered.

Determining Optimal Alignment

Two sequences: X and Y

$|X| = m; |Y| = n$

Allowing gaps, $|X| = |Y| = m+n$

Brute Force

Dynamic Programming

Brute Force

Determine all possible subsequences for X and Y

- 2m+n subsequences for X, 2m+n for Y!

Alignment comparisons

- 2m+n * 2m+n = 2(2(m+n)) = 4m+n comparisons

Quickly becomes impractical

Dynamic Programming

Solve optimization problems by dividing the problem into independent sub problems.

- Sequence alignment has optimal substructure property
- Subproblem: alignment of prefixes of two sequences
- Each subproblem is computed once and stored in a matrix

Steps in Dynamic Programming

- Initialization
- Matrix fill (Scoring)
- Traceback (Alignment)

Dynamic Programming

Sequence #1: GAATTCAGTTA; M = 11

Sequence #2: GGATCGA; N = 7

$s(a_i b_j) = +5$ if $a_i = b_j$ (match score)

$s(a_i b_j) = -3$ if a_i ??b_j (mismatch score)

w = -4 (gap penalty)

Global Alignment

(Needleman-Wunsch Algorithm)

Attempts to align all residues of two sequences

INITIALIZATION: First row and first column set

$S_{i,0} = w * i$

$S_{0,j} = w * j$

Trace Back

(Global Alignment)

Maximum global alignment score = 11 (value in the lower right hand cell).

Traceback begins in position $S_{M,N}$; i.e. the position where both sequences are globally aligned.

At each cell, we look to see where we move next according to the pointers.

$S_{i,j}$ = MAXIMUM /$S_{i-1,j-1}$ + s(a_i, b_j) (match/mismatch in the diagonal), $S_{i,j-1}$ + w (gap in sequence #1),

$S_{i-1,j}$ + w (gap in sequence #2)]

$S_{1,1} = MAX[S_{0,0} + 5, S_{1,0} - 4, S_{0,1} - 4] = MAX[5, -8, -8]$

$S_{1,2} = MAX[S_{0,1} -3, S_{1,1} - 4, S_{0,2} - 4] = MAX[-4 - 3, 5 – 4, -8 – 4] = MAX[-7, 1, -12] = 1$

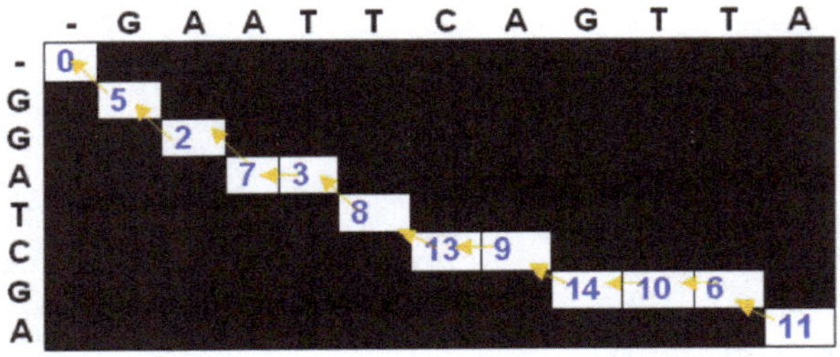


```
G A A T T C A G T T A
|   |   | |   |     |
G G A - T C - G - - A
```

Matrix filled by Needleman-Wunsch Algorithm

Local Alignment

Smith-Waterman Algorithm to obtain highest scoring local match between two sequences

Requires 2 modifications

Negative scores for mismatches

When a value in the score matrix becomes negative, reset it to zero (begin of new alignment).

Values in row 0 and column 0 set to 0

$S_{i,j}$ = MAXIMUM

[
$S_{i-1,j-1}$ + s(a_i, b_j) (match/mismatch in the diagonal),
$S_{i,j-1}$ + w (gap in sequence #1),
$S_{i-1,j}$ + w (gap in sequence #2),
0
]

Example: $S_{1,1} = MAX [S_{0,0} + 5, S_{1,0} - 4, S_{0,1} – 4,0] = MAX[5, -4, -4, 0] = 5.$

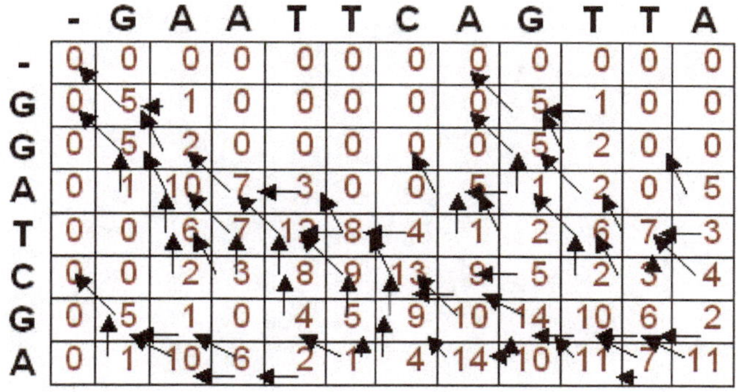

Matrix filled by Smith-Waterman Algorithm

References

BMC Bioinformatics. 2005 Oct 15;6:257 *Statistical Distributions of Optimal Global Alignment Scores of Random Protein Sequences.*

Pang, H., Tang, J., Chen, S.S., Tao, S. 2004. *Conf Proc IEEE Eng Med Biol Soc.* 4:2817-20. *A parallel implementation of the Smith-Waterman algorithm for massive sequences searching.* Liao HY, Yin ML, Cheng Y. 3 Pac Symp Biocomput. 1999:302-13 *Optimizing Smith-Waterman alignments.* Olsen R, Hwa T, Lassig M.

17

Sequence Analysis and Phylogenetic Predictions

S.K. Pradhan

Sequence Alignment

Sequence Alignment is the procedure of comparing two (Pair-wise Alignment) or more (Multiple Sequence Alignment) sequences by searching for a series of individual characters or character patterns that are in the same order in the sequences.

Significance: Discovering functional, structural and evolutionary information in biological sequences.

Types

- Global Alignment (Align across the entire length, need to use gaps-i/d)
- Local Alignment (Align certain region, avoid gaps).

Homology: Two sequences are said to be homologs if they are descended by evolution from the same sequence in the genome of a common ancestor.

Orthologs: It describes sequences in different species that derive from a common ancestor (they may or may not have same function).

Paralogs: It describes sequences within a single species that diverged by gene duplication (orthologs & paralogs are two types of homologs).

Xenologs: Homologs resulting from horizontal gene transfer between two organisms may have different function.

Horizontal Gene Transfer/Lateral Gene Transfer: Any process in which an organism incorporates genetic material from another organism without being the offspring of that organism.

Vertical Gene Transfer: Occurs when an organism receives genetic material from its ancestor e.g its parent or a species from which it is evolved.

Sequence Alignment Algorithm

- Dynamic Programming (slow but optimizing method)
- Heuristic Algorithm (efficient but not as thorough)
- Probabilistic Method

Heuristic algorithm: That is able to produce an acceptable solution to a problem in many practical scenarios, but for which there is no formal proof of its correctness.

Multiple Sequence Alignment

Global Alignment of three or more sequences (DNA/PROTEIN)

The highly conserved residues that define structural and functional domains in protein families can be identified in case of protein sequences and alignment of DNA sequences can assist in finding conserved regulatory patterns in DNA sequence.

MSA Tools

ClustalW2 (NJ), T-Coffee, Muscle, MAFFT, KALIGN, PILEUP (UPGMA) etc.

ClustalW2

It produces biologically meaningful global-MSAs of divergent sequences. It calculates the best match for the selected sequences and lines them up so that the identities, similarities and differences can be seen.

Input Format: FASTA format

Accession Number

ATTGCAA....../MLAFGDR......

Matrix: Blosum-30, PAM-350, Gonnet-250

Default- for DNA : DNA Identity Matrix (positive score for a match and score of -10000 for a mismatch, no substitution is allowed and gap is permitted)

Protein: Gonnet 250

Causes of Gaps

- Single Mutation
- Unequal Crossing Over
- DNA slippage in the replication procedure

- Retrovirus insertions
- Translocation of DNA between chromosomes

Gap Open: Penalty for opening a gap, Default for DNA: 15.0, Protein: 10.0

End Gap: Exclude end gaps, Default: Yes

Gap Ext: Penalty for extending a gap, Default- for DNA : 6.66, Protein: 0.2

Gap Dist: Gap separation penalty, Default: 4

Color

AVFPMILW	RED (small + hydrophobic+ Aromatic)
DE	BLUE (Acidic)
RK	MAGNETA (Basic)
STYHCNGQ	GREEN (Hydroxyl + Amine)
Others	Gray

Symbols denoting the degree of conservation observed in each column.

* Means that the residues or nucleotides in that column are identical in all sequences in the alignment.

: Means that conserved substitutions (substitution of amino acids whose side chains have similar biochemical properties) have been observed.

- Means that semi-conserved substitution are observed

Output Format: aln, gcg, phylip

MSA Vs Phylogenetic Analysis

- After MSA has been found, the number or types of changes in the aligned sequence residues may be used for Phy. Ana.
- Each column in the alignment predicts the mutations that occurred at one site during the evolution of the sequence family

Seq A N . F L S

Seq B N . F - S

Seq C N K Y L S

Seq D N . Y L S

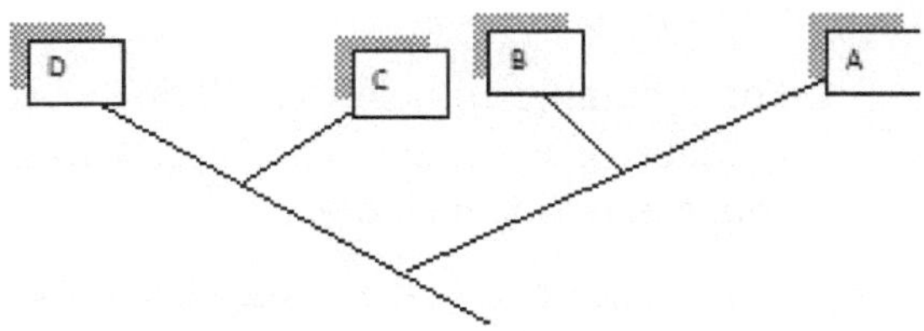

Phylogenetic Prediction

A phylogenetic Analysis of a family of related nucleic acid or protein sequences is a determination of how the family might have been derived during evolution.

Tools: PHYLIP (Phylogenetic inference Package)

(http://evolution.genetics.washington.edu/phylip.html)

PAUP (Phylogenetic Analysis using Parsimony)

(http://www.lms.si.edu/PAUP/)

MEGA 4(Molecular Evolutionary Genetics Analysis)

(http://www.megasoftware.net/)

A phylogram is a branching diagram (tree) that is assumed to be an estimate of a phylogeny. The branch lengths are proportional to the amount of inferred evolutionary change.

A cladogram is a branching diagram (tree) assumed to be an estimate of a phylogeny where the branches are of equal length. Therefore cladograms show common ancestry, but don't indicate the amount of evolutionary time separating taxa

Tree Construction Method

- Maximum Parsimony Method
- Distance Method
- Maximum Likelihood Method
- UPGMA (Unweighted Pair group method using Arithmatic Average)

Maximum Parsimony Method

- Predicts the evolutionary tree that minimises the number of steps required to generate the observed variation in the sequences (also known as minimum evolution method)
- A MSA is required to predict which sequence positions are likely to correspond
- For each aligned position, phylogenetic trees that require the smallest number of evolutionary changes to produce the observed sequence changes are identified
- Finally, those trees that produce the smallest no. of changes overall for all sequence positions are identified
- Used for sequences that are quite similar & small no of sequences

Phylip is a package of programs for inferring phylogenies (Evolutionary tree)

Methods that are available in the package include

Phylip-3.68

- Parsimony Method
- Distance Matrix Method
- Maximum Likelihood Method

Data types that can be handled include

- Molecular sequence
- DNA
- Protein
- Gene frequencies
- Restriction sites and fragment
- Distance Matrices
- Discrete characters

Parsimony Programs

For DNA :

- Dnapars — DNA Parsimony Program
- Dnacomp — DNA Compatibility Program

For Protein

- Protpars — Protein Sequence Parsimony Method

Dnapars

This program carries out unrooted parsimony on DNA sequences.

Assumptions

- Each site evolves independently.
- Different lineages evolve independently.
- The probability of a base substitution at a given site is small over the lengths of time involved in a branch of the phylogeny.

- The expected amounts of change in different branches of the phylogeny do not vary by so much that two changes in a high-rate branch are more probable than one change in a low-rate branch.
- The expected amounts of change do not vary enough among sites that two changes in one site are more probable than one change in another.

Steps

- Open the exe folder of phylip and open dnapars
- Input the '.txt' file in the command prompt (The sequences must either be in the "interleaved" or "sequential" formats)
- Options are selected using an interactive menu for setting
- Type '0' to change terminal type from IBM PC to ANSI
- Type 'Y" to accept the settings
- Then 'outfile' and 'outtree' file will be generated

Dnacomp

This program implements the compatibility method for DNA sequence data.

Assumption

- Each character evolves independently.
- Different lineages evolve independently.
- The ancestral base at each site is unknown.
- The rates of change in most sites over the time spans involved in the divergence of the group are very small.
- A few of the sites have very high rates of change.

Steps

- Open the exe folder of phylip and open dnacomp
- Steps are same as that of Dnapars

Protopars

- This program infers an unrooted phylogeny from protein sequences, using a new method intermediate between the approaches of Eck and Dayhoff (1966) and Fitch (1971).

Assumption

- Changes in different sites are independent.
- Changes in different lineages are independent.
- The probability of a base substitution that changes the amino acid sequence is small over the lengths of time involved in a branch of the phylogeny.
- The probability of a base change that is synonymous is much higher than the probability of a change that is not synonymous.

Steps

- Open exe folder of phylip and open protopars.
- Steps are same as that of Dnapars.

MEGA-4.0.2

The Molecular Evolutionary Genetics Analysis (MEGA) software is a desktop application designed for comparative analysis of homologous gene sequences either from multi gene families or from different species with a special emphasis on inferring evolutionary relationships and patterns of DNA and protein evolution.

Steps

- Install the MEGA 4.0.2 set up
- Open the MEGA from program
- Go to File > open > convert to MEGA Format
- Generate the Phylip format using clustalW2 and select this file to convert it into ".meg" format
- Activate the data file and select the data type
- Data explorer will open, where you can find out the conserved sites, variable sites, Parsimony sites
- Go to Phylogeny to compute the tree selecting various methods also can able to compute the Bootstrap test of phylogeny

References

Sequence and Genome Analysis by David Mount (http://evolution.genetics.washington.edu/phylip.html) (http://www.lms.si.edu/PAUP/) (http://www.megasoftware.net/)

18

Finding Regulatory Motifs in DNA Sequences

S. Padhy

Motif

1. A nucleotide/amino acid sequence pattern having biological significance.
2. Occurs repeatedly, either in same molecule or in many molecules. Found in upstream intergenic region.
3. Functionally important regions of genome (gene) can be recognized by searching such patterns.
4. Used for locating binding sites, regulatory signals, to control gene expression.
5. Identifying potential drug target sites.

Combinatorial Gene Regulation

A micro-array experiment showed that when gene X is knocked out, 20 other genes are not expressed:

- How can one gene have such drastic effects?

Regulatory Proteins

- Gene X encodes regulatory protein, a.k.a. a *transcription factor* (TF).
- The 20 unexpressed genes rely on gene X's TF to induce transcription.
- A single TF may regulate multiple genes.

Regulatory Regions

- Every gene contains a regulatory region (RR) typically stretching 100-1000 upstream of the transcriptional start site.

- Located within the RR are the *Transcription Factor Binding Sites* (TFBS), also known as *motifs*, specific for a given transcription factor.
- TFs influence gene expression by binding to a specific location in the respective gene's regulatory region - TFBS.

Transcription Factor Binding Sites

- A TFBS can be located anywhere within the
- Regulatory Region
- TFBS may vary slightly across different regulatory regions since non-essential bases could mutate.

Identifying Motifs

- Genes are turned on or off by regulatory proteins.
- These proteins bind to upstream regulatory regions of genes to either attract or block an RNA polymerase.
- Regulatory protein (TF) binds to a short DNA sequence called a motif (TFBS).
- So finding the same motif in multiple genes' regulatory regions suggests a regulatory relationship amongst those genes.

Identifying Motifs: Complications

- We do not know the motif sequence.
- We do not know where it is located relative to the genes start.
- Motifs can differ slightly from one gene to the next.
- How to discern it from "random" motifs?

The Motif Finding Problem

- Given a random sample of DNA sequences

ctgatagacgctatctggctatccacgtacgtaggtcctctgtgcgaatctatgcgtttccaaccat
gtactggtgtacatttgatacgtacgtacaccggcaacctgaaacaaacgctcagaaccagaagtgc
aaacgtacgtgcaccctctttcttcgtggctctggccaacgagggctgatgtataagacgaaaatttt
agcctccgatgtaagtcatagctgtaactattacctgccacccctattacatcttacgtacgtataca
ctgttatacaacgcgtcatggcggggtatgcgttttggtcgtcgtacgctcgatcgttaacgtacgtc

- Find the pattern that is implanted in each of the individual sequences, namely, the motif

Additional information

- The hidden sequence is of length 8.
- The pattern is not exactly the same in each array because random point mutations may occur in the sequences.
- The patterns revealed with no mutations:

cctgatagacgctatctggctatccacgtacgtaggtcctctgtgcgaatctatgcgtttccaaccat
gtactggtgtacatttgatacgtacgtacaccggcaacctgaaacaaacgctcagaaccagaagtgc
aaacgtacgtgcaccctctttcttcgtggctctggccaacgagggctgatgtataagacgaaaatttt
agcctccgatgtaagtcatagctgtaactattacctgccacccctattacatcttacgtacgtataca
ctgttatacaacgcgtcatggcggggtatgcgttttggtcgtcgtacgctcgatcgttaacgtacgtc

acgtacgt

Consensus String

- The patterns with 2 point mutations:

cctgatagacgctatctggctatcc**aGgtacTt**aggtcctctgtgcgaatctatgcgtttccaaccat
agtactggtgtacatttgat**CcAtacgt**acaccggcaacctgaaacaaacgctcagaaccagaagtgc
aa**acgtTAgt**gcaccctctttcttcgtggctctggccaacgagggctgatgtataagacgaaaatttt
agcctccgatgtaagtcatagctgtaactattacctgccacccctattacatctt**acgtCcAt**ataca
ctgttatacaacgcgtcatggcggggtatgcgttttggtcgtcgtacgctcgatcgtta**acgtacg**

Defining Motifs

1. To define a motif, lets say we know where the motif starts in the sequence.
2. The motif start positions in their sequences can be represented as ***s*** = (*s*1,*s*2,*s*3,…,*s*t).

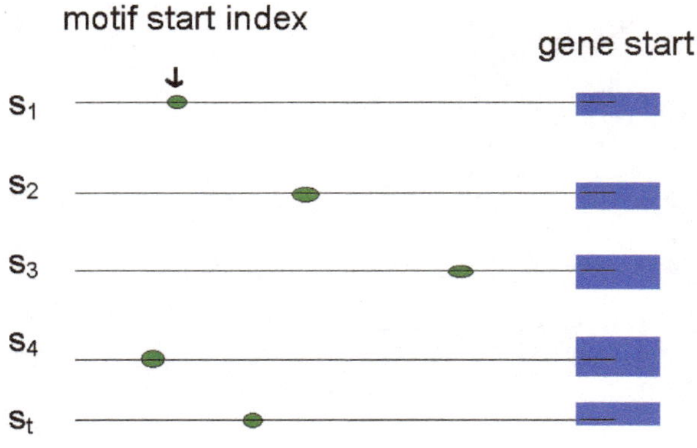

Motifs: Profiles and Consensus

Alignment		a	G	g	t	a	c	T	t
		C	c	A	t	a	c	g	t
		a	c	g	t	T	A	g	t
		a	c	g	t	C	c	A	t
		C	c	g	t	a	c	g	G
Profile	A	3	0	1	0	3	1	1	0
	C	2	4	0	0	1	4	0	0
	G	0	1	4	0	0	0	3	1
	T	0	0	0	5	1	0	1	4
Consensus		A	C	G	T	A	C	G	T

- Line up the patterns by their start indexes
- **s** = (*s*1, *s*2, …, *s*t)
- Construct matrix profile with frequencies of each nucleotide in columns.
- Consensus nucleotide in each position has the highest score in column.

Consensus

- Think of consensus as an "ancestor" motif, from which mutated motifs emerged.
- The *distance* between a real motif and the consensus sequence is generally less than that for two real motifs.

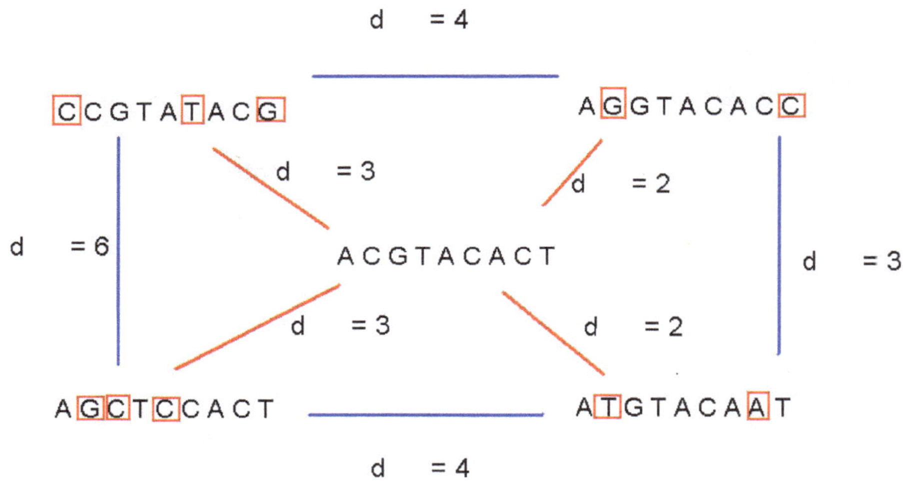

Evaluating Motifs

- We have a guess about the consensus sequence, but how "good" is this consensus?
- Need to introduce a scoring function to compare different guesses and choose the "best" one.

Defining Some Terms

- ***t*** - number of sample DNA sequences.
- ***n*** - length of each DNA sequence.
- ***DNA*** - sample of DNA sequences (*t* x *n* array).
- ***l*** - length of the motif (***l***-mer).
- ***si*** - starting position of an ***l***-mer in sequence *i.*
- s=(*s1, s2,... st*) - array of motif's starting positions.

Scoring matrices

$$\sum_{i=1}^{l} \max_{k \in \{A,T,C,G\}}^{count(k,i)}$$

Score(s, DNA)

```
a G g t a c T t
C c A t a c g t
a c g t T A g t
a c g t C c A t
C c g t a c g G
_______________
A 3 0 1 0 3 1 1 0
C 2 4 0 0 1 4 0 0
G 0 1 4 0 0 0 3 1
T 0 0 0 5 1 0 1 4
_______________
```

Consensus a c g t a c g t

Score 3+4+4+5+3+4+3+4=30

The Motif Finding Problem

- If starting positions **s**=(*s1, s2,... st*) are given, finding consensus is easy even with mutations in the sequences because we can simply construct the profile to find the motif (consensus).
- But… the starting positions **s** are usually not given. How can we find the "best" profile matrix?

The Motif Finding Problem: Formulation

- **Goal:** Given a set of DNA sequences, find a set of ***l***-mers, one from each sequence, that maximizes the consensus score.
- **Input:** A ***t*** x ***n*** matrix of ***DNA***, and ***l***, the length of the pattern to find.
- **Output**: An array of ***t*** starting positions.

  s = (*s1, s2, ... st*) maximizing *Score*(**s**, ***DNA***).

The Motif Finding Problem: Brute Force Solution

- Compute the scores for each possible combination of starting positions s.
- The best score will determine the best profile and the consensus pattern in ***DNA.***
- The goal is to maximize *Score*(**s**,***DNA***) by varying the starting positions *si*, where:

  si = [1, …, ***n-l***+1]

  i = [1, …, ***t***]

Brute Force Motif Search

1. BruteForceMotifSearch(DNA, t, n, l)
2. best Score ? 0
3. for each s=(s1,s2 , . . ., st) from (1,1 . . . 1) to (n-l+1, . . ., n-l+1)
4. if (Score (s,DNA) > bestScore)
5. bestScore ? score(s, DNA)
6. bestMotif ? (s1,s2 , . . . , st)
7. return bestMotif

Running Time of Brute Force Motif Search

- Varying $(n - l + 1)$ positions in each of t sequences, we're looking at $(n - l + 1)t$ sets of starting positions.
- For each set of starting positions, the scoring function makes l operations, so complexity is $l\,(n - l + 1)t = O(l\,nt)$.
- That means that for $t = 8$, $n = 1000$, $l = 10$ we must perform approximately 1020 computations – it will take billions years.

CONSENSUS: Greedy Motif Search

- Find two closest l-mers in sequences 1 and 2 and forms *0 2x l alignment matrix* with *Score (***s***,2,DNA)*.
- At each of the following *t-2* iterations CONSENSUS finds a "best" *l*-mer in sequence *i* from the perspective of the already constructed *(i-1) x l* alignment matrix for the first *(i-1)* sequences.
- In other words, it finds an *l*-mer in sequence *i* maximizing *Score(***s***,i,DNA)*, under the assumption that the first *(i-1) l*-mers have been already chosen.
- CONSENSUS sacrifices optimal solution for speed: in fact the bulk of the time is actually spent locating the first 2 *l*-mers.

Some Motif Finding Programs

1. **CONSENSUS**

   Hertz, Stromo (1989)

2. **GibbsDNA**

   Lawrence et al (1993)

3. **MEME**

   *Bailey, Elkan (*1995)

4. **RandomProjections**

   Buhler, Tompa (2002)

5. **MITRA**

   Eskin, Pevzner (2002)

6. **Pattern Branching** Price,

   Pevzner (2003)

19

Algorithms and Tools for Sequence Alignment

S. Padhy

Objectives

- Creation and use of large databases of Biological information to augment traditional laboratory based biology.
- Analysis and interpretation of genomic and structural data using computational techniques from Mathematics, Statistics and Computer Science.
- Explain the process and mechanism of evolution by comparing the Genomes of different species.
- Development of new methods and technologies for understanding and diagnosis of genetic diseases and design of drugs for their treatment.

Outline

- Basics of Molecular Biology
  - DNA, RNA, Transcription, Translation, Genetic Code, GENE, GENOME, Protein Structure etc.
- Formulation of Biological problems as Computational problems Overview of Algorithms and techniques for :
- Longest Common Substring
- Pairwise sequence Alignment (Global & Local)
- Multiple sequence alignment
- Fragment Assembly
- Gene finding

- Genome Rearrangement
- Computational Drug Discovery Current Topics

Computational Formulation of Biological Problems

Identification of remains of military personnel

A small interval of mitochondrial DNA of deceased person is sequenced and matched against a database of strings determined earlier

- To check if the string is a substring of one of the strings in the database. or more realistically.
- To compute the longest common substring found in the new string and one of the strings in the database the knowledge of which will narrow the possibility of the identity of the person.

Longest Common Substring Problem

- Given a set of strings { *Si*} and another string S, find the largest substring in S that is also in some string *Si*.
- Theorem:
- The longest common substring of S and some strings of { *Si*} can be found in linear time using a generalized **Suffix Tree**.

Remark

In 1970 Knuth conjectured that a linear time algorithm for this problem would be impossible

Computational Formulation of Biological Problems

Recognizing DNA Contamination

- Contamination during cloning from
- Vector
- Host Organism
- Other ways
- The announcement of extraction of DNA from Dinosaur bones has been refuted later (1995) through DNA database searching. The extracted DNA sequence is found to be similar to that of human being than to bird or crocodile suggesting that the DNA in hand was from human contamination. (Wasted sequencing)

- DNA sequences of many of the possible contaminants are known. (cloning vectors, PCR primers, Host organisms like yeast)

DNA Contamination Problem

- Given a string *S1 (newly sequenced)* and a known string *S2(Contaminant)* find all substrings of *S2* that occur in *S1* that are larger than some given length.
- This problem can be solved using *Suffix Trees* in linear time.

Pair Wise Sequence Alignment

When a new gnome is sequenced the first analysis performed is to determine their function which requires biological experiments of high order creativity and complexity.

Pair-Wise Sequence Alignment

FACT: In Biomolecular sequences *(DNA, RNA or Amino acid sequences)* high sequence similarity usually implies significant functional or structural similarity. *(converse is not true)*

" We didn't know it at the time, but we found out every thing in life is so similar, that the same gene that work in flies are the ones that work in humans."

- Eric Wieschaus, Co-Winner 1995 Nobel Prize in Medicine for work of genetics of Drosophila Development
- Using the above fact redundancy can be reduced. Biological function of a gene can be inferred by sequence similarity search on DNA/Protein databases using Pair-wise sequence alignment algorithms.

  – Given two sequences *(DNA/Protein)* find the best alignment between them

Biological Discoveries Through Similarity Search

1. Cancer causing g-sys oncogene of Simian sarcoma virus was compared to a database of all known genes.

   It was discovered that oncogene matched a normal gene involved in growth and development conclusion was that cancer might be caused by a normal growth gene being switched on at the wrong time. (Doolittle *et al.*, 1983)
2. Cystic fibrosis (disease associated with recurrent respiratory infections and abnormal secretions) gene was compared with a database of known genes.

- It was found to be similar to ATP binding proteins which are known to work as channels for the transport of ions across the membrane.
- This discovery led to efficient medical diagnostics and a promise for potential therapy for Cystic Fibrosis (1989).

Computational Formulation of Biological Problems

- Largest Common Subsequence Problem
  - Given two strings S1 and S2, find their largest common subsequence

Multiple Sequence Alignment

- Multiple Sclerosis (MS)
  - An autoimmune disease in which *myelin sheath* encasing nerve cells is attacked by the immune system disrupting the normal transmission of the signals along the nerve. *T-cells* incorrectly identify the protein on the surface of *myelin sheath* as foreign leading to their destruction.

    (Reported in New York Times, May 3, 1995)

Multiple Sequence Alignment

- Conjecture
  - The myelin sheath proteins identified by T-cells were similar to viral/ bacterial sheath proteins from an earlier infection.
  - Later it was confirmed by a search in the protein databases of bacteria and viruses. *(About 100 similarities were found)*
- By Multiple Sequence comparison one might find what features of myelin surface proteins are used by T-cells to mistakenly identify it as foreign

Computational Formulation of Biological Problems Fragment Assembly of DNA *(Sequencing a large GNOME)*

- Difficult to sequence contiguous stretches of more than 700 bases
- Identical copies of the original strand is created by cloning
- Each copy is cut into small fragments which are then sequenced
- Assembly the whole sequence

Fragment Assembly of DNA

- Shortest Common Superstring Problem
  - Given a collection F of strings find a shortest possible string *S* such that for every f Î F, *S* is a super-string of f

Example:

F = { A C T, C T A, A G T}

S = A C T A G T

Definition: The Overlap Multigraph OM(F) of a collection F is the directed, weighted Multigraph defined as follows.

– The set V of nodes is F itself. A directed edge from a Î F to a different fragment b Î F with weight t ³ 0 exists if the suffix of a with t characters is a prefix of b.

Remark: Self loops are not allowed but the edges with zero weight are allowed

No of 0-wt. edges: *n(n-1)* (Any two strings share a common part with zero characters)

Overlap Multigraph

P determines the following alignment

P determines the following alignment

T A C G A - - - - - - - - - - - - - - - - -

- - - - A C C C - - - - - - - - - - - - - -

- - - - - - - - C T A A A G - - - - - - - -

- - - - - - - - - - - - - - G A C A - - - -

S(P): T A C G A C C C T A A A G A C A

If we get an Hamiltonian path P, S(P) is a common superstring of

A = F and we get, | S(P)| = ||F || - w(P)

Thus minimizing S(P) is equivalent to maximizing |w(P)|

- Therefore, we look for a Hamiltonian path of maximum weight in a directed multigraph. This problem is NP- Complete

Overlap Multigraph

- Definition: The graph obtained by considering only greatest weight edges between two nodes of OMG(F) is called the overlap graph of *Fi* and is denoted on OG(F)

This greedy strategy does not always yield optimal results.

Ex: If f=TGCAT,g=ATGC,h=GCC

Then w(f,g)=2,w(g,f)=3,w(g,h)=2,

W(f,h)=0.

The Algorithm yields a Hamiltonian path

gfh of weight 3 where as the Hamiltonian

Path fgh is of weight 4.

THE NP- COMPLETENESS IS APPLICABLE TO AN ARBITRARY

COLLECTION OF FRAGMENTS . HOWEVER FOR A SPECIAL TYPE OF FRAGMENTS THERE EXIST A CORRECT AND EFFCIENT ALGORITHM.

Let S be a string over the alphabet { A,C,G,T}.

An interval of S is [i , j], 1<= i<=j<=|S|

A collection A of intervals of S is a sampling of S

The sampling A covers S if for any I there is an interval of S Which contains I.

A is connected at level t if for every pair of intervals b and c in A There is a series of intemediate intervals such that the length of Overlap of two such consecutive intervals is >= t. Ex. S=ATCGCGATGA ,|S| =10

A ={ [1,5] ,[3,6] ,[4,10]} covers S and is connected at level t.

S[A]= { ATCGC ,CGCG,GCGATGA] } .

A is subinterval free if there are no two intervals such that one is Contained in the other.

A repeat in S occurs if two different intervals of S give rise to the same sub string.

Theorem

Let A be a subinterval free, connected at level t Sampling of S that covers S and F=S(A). If S has no repeats of size t or larger then the graph OG(F,t) is acyclic and has a unique Hamiltonian Path P and S(P)=S.

Algorithm

Since the graph has exactly one source (being Acyclic it has at least one source and for having a Hamiltonian path it can have at most one source)

It must be the starting point of the Hamiltonian path.

Remove this source from the graph, the remaining Graph is still acyclic and has a Hamiltonian path Keep removing sources until none remains. The vertices removed in the above order determines

The Hamiltonian path P from which the required

Superstring S=S(P) is obtained.

Computational Formulation of Biological Problems

Genome Rearrangement

- [A chromosome is defined as a sequence of genes, while genome is defined as a set of chromosomes]
- Waardenburg's syndrome (hearing loss and pigmentary dysplasia) gene is found in chromosome 2 (human)
- Through breeding splotch gene was mapped to mouse chromosome 2
- During evolution blocks of genetic material remain intact even as multiple chromosomal rearrangements occur
- Chromosome 2 in humans consists of mouse DNA residing in mouse Ch. 1, 2, 6, 8, 11, 12 and 17
- Mapping a gene in mice gives a clue to the location of a related human gene
- Nadeau and Taylor,1984: 178 ± 39 genomic rearrangements have happened since the divergence of human and mouse 80 million years ago

This Greedy strategy does not always

yield optimal result.

Ex. : GCC(h)

ATGC(g) ——2———(f)TGCAT

References

D. Gusfield-Algorithms on Strings, Trees and Sequences-Computer Science and Computational Biology-Cambrigde Univ. Press-1997.

N.C. Jones, P.A. Pevzner-An Introduction to Bioinformatics Algorithms-MIT press-2004 P.A. Pevzner-Computational Molecular Biology-An Algoritmic Approach-MIT 2000.

20

Applications of Artificial Neural Network in Biological Research

S. K. Pradhan

ANN is a Mathematical model or Computational model based on biological neural networks. It consists of an interconnected group of artificial neurons and processes information using a connectionist approach to computation. Neural Networks are non-linear statistical data modeling tools. That can be used to model complex relationships between inputs and outputs or to find patterns in data.

Why ANN?

- Neural Networks derive meaning from complicated or imprecise data, can be used to extract patterns and detect trends that are too complex to be noticed by either humans or other computer techniques.
- **Adaptive Learning:** An ability to learn how to do tasks based on the data given for training or initial experience.
- **Self Organization:** An ANN can create its own organization or representation of the information it receives during learning time.

Architecture

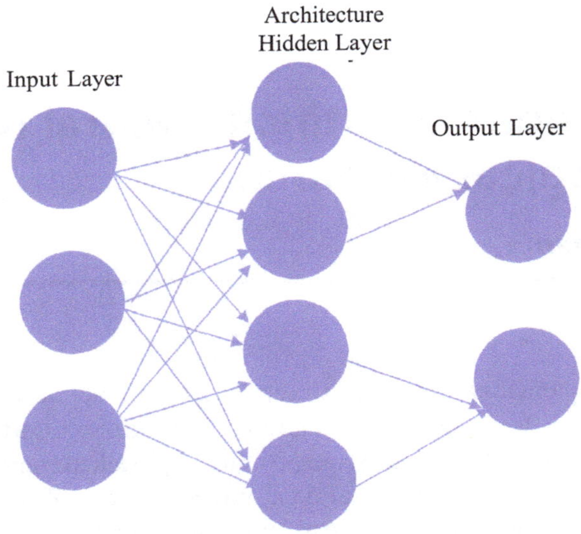

- A NN consists of interconnected Units, often arranged in layers
- Units are also known as a neuron or node is the main processing element of the neural network. It receives a set of input signals and combined with an internal function, converts the input signals to an output signals
- The configuration of these units is known as the architecture and can vary widely depending on the application for which it is used
- Simplest NN having only two layers; one input layer and one output layer-Known as Perceptrons; Multilayer perceptron adds a number of hidden layers of units in between input & output layer

Neural Networks Vs Conventational Computers

Conventional computers use an algorithm approach i.e the computer follows a set of instructions in order to solve a problem. Unless the specific steps that the computer needs to follow are known, the compute cann't solve the problem. Neural Networks process information in a similar way the Human Brain does. The network is composed of a large number of highly interconnected processing elements(Neurons) working in parallel to solve a specific problem.

The network in ANN

The word network in the term "ANN" because the function f(x) is defined sa a composition of other functions gi(x), which can further be defined as a composition of other function.

F(x)=K(“wigi(x))

K=some pre defined function

g=(g1,g2,g3........gn)

Supervised Learning

Supervised learning is the process of learning relationships between input and output data.In this type of learning, input data is passed to the input layer,propagated through the units and weights of the neural network and the response of the network is compared to the required response which is dictated by the output data. The simplest learning rule is that of Rosenblatt’s perceptron

Perceptron learning rule

1. Initialize the network weights and unit thresholds to some small random values.
2. Present the input data (I1, I2, I3.........In) and the desired ouput(o) data to the network.
3. Calculate the output from the network using the expression:F(s)[“ Wk(t)Ik(t)]
4. Adapt the weights
   - If correct Wk(t+1)=Wk(t)
   - If output is 0 and should be 1 Wk(t+1)=Wk(t)+Ik(t).

Where Wk(t) is the weight value at time t,Wk(t+1) is the weight value at time(t+1)(after updating) and F(s) is the function used to compute the output in the network.

Execute Network

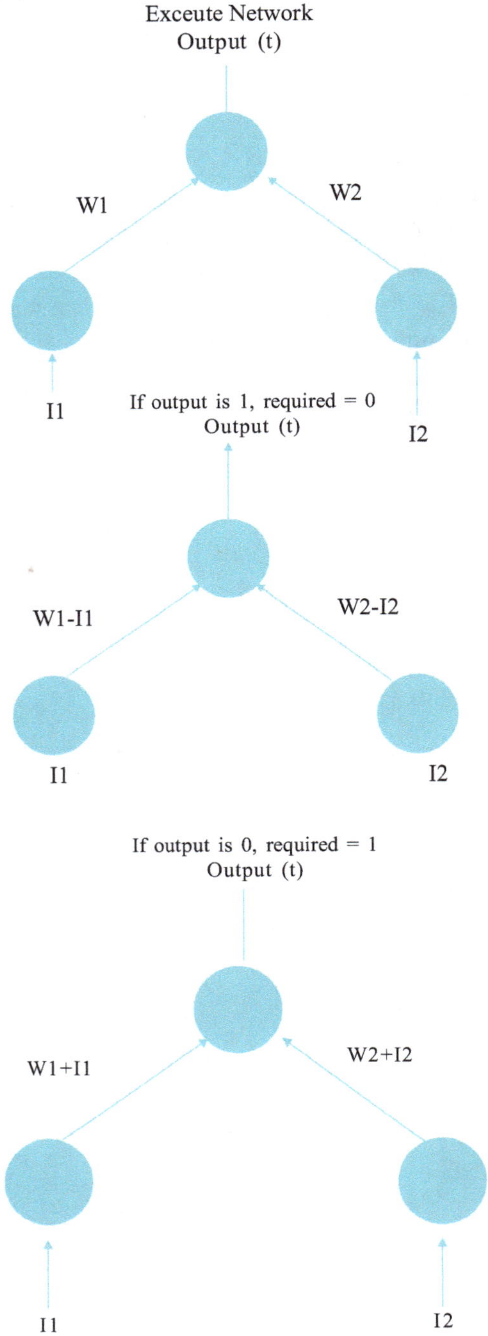

Multi-layer Perceptron Learning Rule

For larger architectures with a hidden layer a more complex learning rule is required .The Multi-layer Perceptron often uses the sigmoid function , therefore when the difference between the actual and desired ouput patterns is calculated the weights must be changed in accordance with the derivative of this function , as opposed to the simple increment/decrement approach when considering the step function .Also because more than one layer of weights is considered, the learning rule take into account this fact and contain some method of backpropagating the error through the network.The multi-layer perceptron learning rule therefore implements the sigmoid derivative and back propagation to allow it to learn.

1. Initialize the network weights and unit thresholds to some small random values.
2. Present the input data (I1,I2,I3.........In) and the desired ouput(o) data to the network.
3. Calculate the output from the network using the expression F(s)[" Wk(t)Ik(t)] for each layer in the network. The output from the final layer is the vector of output values.
4. Adapt the weights, startning from the output layer and moving backwards using the equation:

   Wij(t+1)=Wij(t)+äpj Opj

Where, Opj is the output of node j for pattern P and äpj is defined as: For the output layer: äpj=Copj(1-Opj)/(dpj-Opj)

For the hidden layer: äpj=Copj(1-Opj)" äpk-Wjk

Where 'd' is the desired output, o is the actual output and C is a constant used in the sigmoid function.

Types of Application

1. Feed forward Neural Network
2. Radial basis function(RBF) Network
3. Kohonen self organisation Network
4. Recurrent Network
5. Simple Recurrent Network
6. Hope field Network

7. Echo state Network
8. Stochastic Neural Network

Application

- Classification and dimensionality reduction of gene expression data
- Identifying Protein subcellular Location
- TATA BOX Motifs Identification in Eukaryotes
- Promoter Recognition in Genomic DNA
- Other non-biological Application
- Image Processing
- Speech Recognition
- Character Recognition
- Natural Language Processor
- Pattern Recognition
- Weather forecasting

Some ANN Based Programmes

NNPP, Promoter 20, SPNN, SPNN2, McPromoter, Dragon Promoter Finder, Dragon Gene Start Finder (Promoter Recognition programme)

Kohonen Neural network is used for identification of sub cellular location of protein.

References

S.Rajesekeran and G.A. Vijoylaxmi Pai. Neural Network, Fuzzy logic and Genetic Algorithms PHI-2003.

E. Keedorell and A. Neroyenen, 2005. Intelligent Bioinformatics. John Wiley.

L. Worg. 2004. The Practical Bioinformatics. World Scientific Publishing.

21

Homology Modeling using Modeller 9v6

S.N. Rath

Insights into the three-dimensional (3D) structure of a protein are of great assistance when we plan experiments to understand protein function and in assistance of the drug design process.

Homology modeling

- Homology modeling consists of the extrapolation of the structure for a new (target) sequence from the known 3D-structure (X-ray crystallographic or NMR) of related family members (templates).
- Homology modeling is for those targets that have homologous proteins with known structure (>30% identity)
- Homology modeling is for easy targets.
- Homology modeling treats the template in an alignment as a sequence and only sequence homology is used for prediction.
- When the sequence identity between the target and template sequence is low (<25%), homology modeling may not produce a significance result.

Steps involved in homology modeling

- Identification of modeling template.
- Alignment of the target sequence with the template sequence.
- Building the model.
- Evaluation and Refinement of the Structure.

Homology modeling with MODELLER 9v6

MODELLER is a sophisticated program that attempts to automate the homology modeling process. It is freely available for academically purpose. It is capable of using multiple sequence alignments, and can read sequences and alignments in a number of standard database formats. The main functions of the program include a distance geometry, local energy minimisation, and a molecular dynamics. Given a sequence and template structure(s), it can make an alignment and decide which parts of a template structure are probably conserved in the sequence. It is also possible to specify an alignment, which MODELLER uses to generate appropriate spatial and torsional restraints. The distance geometry algorithm then uses the constraint set to construct rough models for the sequence. These are refined using energy minimization and simulated annealing, by combining the constraints with a simple molecular mechanics energy function. MODELLER accomplishes these steps using scoring functions that balance the physical requirements of the modeled sequence against the geometric relations obtained from the template structures. This is essential for its loop modeling - an additional function which uses knowledge based scoring functions that integrate observations on how composition, location, and end to end distance relate to observed loop conformations. The latest version of MODELLER is MODELLER 9v6.

Preparing input files

The sample input files in this tutorial can be found in the examples / automodel directory of the MODELLER distribution. There are three kinds of input files: Protein Data Bank atom files with coordinates for the template structures, the alignment file with the alignment of the template structures with the target sequence, and MODELLER commands in a script file that instruct MODELLER what to do.

Preparing alignment file

- Take protein sequence of your interest and search it in NCBI and save its fasta sequence as in following example

  TARGET

  >Q865F8|AHSP_BOVIN Alpha-hemoglobin-stabilizing protein - Bos taurus

  ALIQTNKDLISKGIKEFNILLNQQVFSDPAISEEAMVTVVNDWVSF YIN

  Y YK KQLSGEQDEQ DKALQEFRQELNTLSASFLDKYRNFLKSS

- Copy your target protein sequence and paste in NCBI blast-p page by choosing.

  PDB database in BLASTP option.

- Get the structural sequence based on degree of similarity and note down pdb id basing on 3 parameters identity more than 40 %,x-ray crystallographic structure, resolution<3Angstrom,R-value<0.5,for this case the template is 1y01
- Copy the sequence which shows highest similarity with your target protein sequence and paste it in same note pad in which your target sequence was pasted in first step. Your note pad should be like below :

  >1y01

  LLKANKDLISAGLKEFSVLLNQQVFNDALVSEEDMVTVVE DWMNF

  YIN YYRQQVTGEPQERDKALQELRQELNTLANPFLAKYRDFLKS

  >target

  MALIQTNKDLISKGIKEFNILLNQQVFSDPAISEEAMVTVVNDWVSFYINY

  YKKQLSGEQDEQDKAL QEFRQELNTLSASFLDKYRNFLKSS

- Save the file where you wish to store then go to Clustalx Open Clustalx software choose upload sequence and load your file which you have saved at step 5 on note pad.
- After uploading click alignment button then click output format options where a Sub-window displayed on your monitor and adjust the parameters.

  Select-parameter out–ON

  Select-PIR format-ON Remove Clustal W format. Then close the window.

- Again click Alignment button and click "do complete alignment button".
- Now your output files will be created in your folder (from there you have been uploaded your sequence file to Clustalx).
- Open the out put file, which has *. pir extension. Open that file through word pad and save as filename with ali extension (*.ali). (* Indicates file name). File should be saved within double quotes ("*.ali")
- Then do modifications in *.ali file by observing standard file, which is given below.
- **Ali File**

  >P1;1y01

structureX:1y01:3:A:91::ALPHA-HEMOGLOBIN STABILIZING PROTEIN:HOMO SAPIENS:2.80: 0.273

—

LLKANKDLISAGLKEFSVLLNQQVFNDALVSEEDMVTVVED WMNF

YINYYRQQVTGEPQERDKALQELRQELNTLANPFLAKYR DFLKS-*

>P1;target

sequence:target:1::102::Alpha-hemoglobin-stabilizing protein:Bos taurus:
2.80:0.273

MALIQTNKDLISKGIKEFNILLNQQVFSDPAISEEAMVTVVNDWVSFYIN

YYKKQLSGEQDEQDKALQEFRQELNTLSASFLDKYRNFLKSS
*

Preparing Atom file

- Take the pdb id, which has got match with your target protein, submit to pdb database (Protein Data bank)
- The database (Access at http://www.rcsb.org)
- Structural data will be opened and choose explore button then download and display option and choose pdb format x in download options for downloading.
- File name should be in atm (*.atm) or pdb (*.pdb) or ent (*.ent) format.

Preparing Script files

- Write the standard format of script file, which is given below
- Then fill the fields required by the modeler.
- Then save the file by giving the file name with the extension of *.py.
- Copy all three (*.ali, *.py & *.atm) and paste it in C:\mod9v6\bin.
- Then open DOS program go to specified path to be given here. (C:\Program Files\Modeller9v6\bin)
- Type command "mod9v6 *.py (* indicates filename).

- Out put file will be given created automatically by Modeller in bin directory approximately 3 min of analysis.
- Copy that output file and paste it in your folder then save the same file as *.pdb

Py File (script file)

```
# Homology modeling by the automodel class
from modeller import * # Load standard Modeller classes from
modeller.automodel import*# Load the automodel class log.verbose() #
request verbose output
env = environ() # create a new MODELLER environment to build this
model in
# directories for input atom files
env.io.atom_files_directory = './:../atom_files'
a = automodel (env,
alnfile = 'align.ali',     # alignment filename

    knowns  = '1y01',              # codes of the templates
    sequence = 'target')           # code of the target
    a.starting_model= 1            # index of the first model
    a.ending_model = 1             # index of the last model
                                   # (determines how many models to
                                     calculate)
    a.make()                       # do the actual homology modeling
```

Visualize your 3D model created by you in Rasmol, Swiss PDB viewer

References

Eswar, N., Marti-Renom, M. A., Webb., Madhusudhan, M. S,. Eramian, D., Shen, M., Pieper,U., Sali, A., 2006. Comparative Protein Structure Modeling With MODELLER. Current Protocols in Bioinformatics, Supplement 15, 5.6.1-5.6.30.

http://salilab.org/modeller/manual/node7.html

http://salilab.org/modeller/manual/node11.html

http://salilab.org/modeller/manual/node14.htm

http://salilab.org/modeller/manual/node271.htm

22

Evolutionary Perspective of Codon Usage Analysis in Human Disease Genes

B. P. Barik

The genetic code uses 64 codons to represent the 20 standard amino acids and the translation termination signal. Each codon is recognised by a subset of a cell's transfer ribonucleotide acid molecules (tRNAs) and with the exception of a few codons that have been reassigned in some lineages (Osawa and Jukes 1989; Osawa *et al.*, 1990) the genetic code is remarkably conserved, although it is still in a state of evolution (Osawa *et al.*,1992). The genetic code uses 64 codons to represent the 20 standard amino acids and the translation termination signal. Each codon is recognised by a subset of a cell's transfer ribonucleotide acid molecules (tRNAs) and with the exception of a few codons that have been reassigned in some lineages (Osawa and Jukes 1989; Osawa *et al.*, 1990) the genetic code is remarkably conserved, although it is still in a state of evolution (Osawa *et al.*, 1992).

The evidence that natural selection could influence silent changes (Grantham *et al.*, 1981; Grantham *et al.*, 1980b; Ikemura 1985; Kimura 1983), suggested that in some (perhaps many) genes in some (perhaps many) species, silent sites were not neutral (Sharp *et al.*, 1993). The corollary of this is that some synonymous substitutions are effectively neutral and probably accumulate at frequencies approaching the mutation rate (Ikemura 1981a; Ikemura 1981b; Ikemura 1985; Ochman and Wilson 1987b). Analysis of codon usage can infer both the nature and strengths of some of the selective forces to which the organism has been exposed (Sharp and Cowe 1991). They can reveal the rates and patterns of silent site evolution and allow the investigation of how natural selection selects between mutations that (presumably) cause very small differences in fitness (Akashi 1995). Weak selection allows non-adaptive processes to be evident and with a large number of sites under broadly similar constraints, it is possible to perform the quantitative analysis of data. With the identification of advantageous codons it is possible to predict the relative advantage and disadvantage of alternative sequences and perhaps the prediction of a completely optimal sequence (Akashi, 1995).

Identification of novel genes associated with human diseases is one of the critical issues in medical research. After completion of human genome project and continuously growing updated database of genes and disorders, evolutionary features in terms of codon usage will unravel the complex relationship between genes and disease.

Literature Survey

The exponential increase in the volume of sequence information during the early 1980s facilitated for the first time detailed statistical analyses of codon usage. Multivariate analysis techniques were applied to the analysis of the codon usage in mammalian, viral, bacteriophage, bacterial, mitochondrial and lower eukaryote genes (Grantham *et al*., 1980a; Grantham *et al*., 1981; Grantham *et al*., 1980b). The results of Grantham and co-workers demonstrated that genes could be grouped based on their codon usage and that these groups agreed broadly with taxonomic groupings. Consequently, they proposed the Genome Theory, which was "that the codon usage pattern of a genome was a specific characteristic of an organism". Compilations of codon usage information have confirmed broadly this organism specific codon choice pattern (Aota *et al*., 1988; Aota and Ikemura 1986). It was suggested that this variation in codon usage might be correlated with variation in tRNA abundance (Grantham *et al.* 1980b), and that this might "modulate" gene expression (Grantham *et al*., 1981). The non-random usage of codons and variation in codon usage between species suggested some selective constraint on codon choice. The codon usage of thirteen strongly and sixteen weakly expressed *E. coli* genes was examined, again using a multivariate analysis technique, and was found to have a marked variation in codon usage (Grantham *et al*., 1981).

A modulation of the coding strategy according to expression was proposed, such that codons found in abundant mRNAs were under selection for optimal codon-anticodon pairing (Grantham *et al*., 1981). A later codon usage analysis of 83 *E. coli* genes found that variation in codon usage was dependent on translation levels, and that codon usage of abundant protein genes could be distinguished from that of other *E. coli* genes (Gouy and Gautier 1982). Genes with a high protein copy number used a higher frequency of intermediate energy codons and codons that required fewer tRNA discriminations per elongation cycle (Gouy and Gautier 1982). The distribution of codon bias in *E. coli* was initially reported as bimodal (Blake and Hinds 1984), but it is now accepted that the distribution is unimodal, which presumably reflects a continuum of expression levels (Holm 1986; Ikemura 1985; Sharp and Li 1987a). The distribution of codon bias in *S. cerevisiae* as calculated by the codon bias index and cluster analysis of codon usage, was also described as bimodal (Sharp and Li 1987a; Sharp *et al.*, 1986). The original clear distinction between highly and lowly expressed genes was not as apparent in a later analysis but variation in the usage of optimal codons remained the main source of heterogeneity among *S. cerevisiae* genes (Sharp and Cowe 1991). Codon usage differs between species not only in the selection of codons but in the degree of bias. *B. subtilis* has less biased codon usage than *E. coli*, presumably reflecting a weaker selection, perhaps due to its different environment affecting its effective population size (Moszer *et al.*, 1995; Ogasawara 1985; Shields and Sharp 1987). On the other hand codon bias in *S. cerevisiae* is much stronger than in *E. coli* (Sharp *et al.*, 1993). Difference in codon bias of homologous genes does not necessarily imply a difference in the expression levels, but rather, it suggests that the effectiveness of the selective pressures on codon usage are not the same.

It has been proposed that synonymous codon usage of human tissue-specific genes was under selective pressure to modulate the expression of proteins by codon-mediated translational control (Plotkin, J.B., H. Robins, and A.J.Levine; Proc. Natl. Acad. Sci. USA 101:12588-12598, 2004). And it has been confirmed that synonymous codon differs significantly between the tissues although the effect is very weak. Variations of synonymous codon usage between tissue-specific genes expressed in different tissues are due to regional variations of substitution patterns and not to translational selection (Semon, Lobry and Duret; Mol. Biol.23 (3):523-529.2006.

Motivation

The human genome has an extremely complex tissue expression profile. Some genes are expressed only in certain tissues during specific times, while others are constitutively and ubiquitously expressed. For the latter genes, they are presumed to be necessary for the most fundamental cellular physiological processes and are referred as housekeeping genes. Housekeeping genes have been studied by many researchers and some interesting observations have been reported. For example, Zhang and Li found that housekeeping genes evolved more slowly than tissue-specific genes. Eisenberg and Levanon found that housekeeping genes were compact in their coding lengths, which could be the result of higher selective pressure.

Based on the unique properties of the ubiquitously expressed human genes (UEHGs), it may be believed that they are suitable candidates for essential genes. Although this hypothesis is intuitive and sounds reasonable, serious efforts are required to collect supportive evidence on a systematic level.

Ubiquitously expressed human genes (UEHGs) are very likely to contain a large proportion of essential genes. By performing a three-way feature comparison of UEHGs (presumed essential genes), disease genes, and the rest of human genes (referred as other genes), we may be able to understand that how they are different in many aspects such as the evolutionary conservation rates, DNA coding lengths, gene functions, etc.

Codon usage analysis may reveal comparative account of housekeeping genes, disease genes and other genes.

Compiling Lists of Disease Genes and UEHGS List

The list of disease genes were obtained from OMIM. Over 4,000 records were listed in the morbid map (March 4,2008) and entries with known sequence (OMIM ID marked with*), with known sequence and phenotype (OMIM ID marked with #), and with phenotype description, molecular basis known (OMIM ID marked with +) are considered.

The list of housekeeping genes is derived from "Trends in Genetics"19,362-365 (2003). Ubiquitously expressed genes can also be obtained from the gene expression experiment results.

Other human disease related gene centered information available elsewhere may be collected for consideration in this study.

Systematic differences in synonymous codon usage between genes selectively expressed in housekeeping genes, disease genes and other genes may reveal the expression pattern and evolutionary preservation of codon usage in humans.

Availability of Software for Codon Usage Analysis

There is a surprising scarcity of software designed to aid the analysis of codon usage. There are programs available that tabulate species codon usage directly from the public nucleic acid databases (GenBank) (Nakamura *et al.*, 1996). Some programs can calculate particular codon usage indices such as: the effective number of codons (Nc) (Wright, 1990); the codon bias index (CBI) (Bennetzen and Hall, 1982); or more general indices of codon bias (Goldman *et al.*, 1995; Krishnaswamy and Shanmugasundaram 1995; Rodriguezbelmonte *et al.*, 1996). However these programs use different input formats and restrict themselves to the calculation of a single index. One program that has integrated several codon usage indices with other measures of codon composition was CODONS (Lloyd and Sharp, 1992), written in FORTRAN 77 (ANSI 1978). CODONS served as the inspiration for the package described herein. GCG (GCG 1994), the most commonly used sequence analysis package includes programs for codon usage analysis. CODONFREQUENCY tabulates codon usage, CODONP

REFERENCE predicts protein-encoding regions using the tabulated codon usage, and CORRESPOND utilises codon usage patterns to generate a matrix of Euclidean Manhattan distances between genes. The formulae used by CORRESPOND (Equation 2-11), are related to those used during the preliminary stages of a correspondence analysis. However, the program name is somewhat misleading as CORRESPOND stops at this stage – equivalent to an early stage of a correspondence analysis (i.e. the generation of a matrix of distances), and does not actually perform a true analysis of correspondences. Recently a world wide web based resource (URL http://acnuc.univ-lyon1.fr/mva/coa.html) was designed to facilitate the multivariate statistical analysis of codon usage (Perriere and Thioulouse, 1996). This resource integrates WWW-Query (URL http://acnuc.univlyon1. fr/start.html) (Perriere and Gouy, 1996), the web interface of the ACNUC database interrogation and sequence retrieval software (Gouy *et al.* 1985), with NetMul (URL http://biomserv.univ-lyon1.fr/base.html) a web interface for multivariate analysis methods (Thioulouse and Chevenet 1996). NetMul is a subset of the ADE (Analysis of Environmental Data) package (Thioulouse *et al.* 1995). These interfaces are integrated such that a group of sequences (i.e. a dataset) can be created using WWW-Query, the codon usage (or amino-acid usage) tabulated, and a MVA of the data performed using NetMul. The positions of the genes along the trends identified by the correspondence analysis are then displayed as a scatter plot (a GIF file which includes HTML map coordinates) (Thioulouse, 1996). There are limitations to using a web-based analysis system. As with all web-based resources, there is the problem of overall lack of bandwidth on the Internet. The interpretation of results from a MVA of codon or amino acid usage is difficult without reference to other properties that can often be used to quantify the trends. Sequence annotation is often incorrect and in many cases inconsistent, thus automatically generated datasets can include data of poor quality, which can severely distort the results of an analysis. All datasets must be carefully inspected. This web resource, in common with the other programs previously mentioned, does not attempt to identify potentially optimal codons.

Codon W

CodonW was designed to simplify the analysis of codon usage, by integrating codon usage indices with multivariate statistical analysis (i.e. correspondence analysis) into a single program. It had to be simple to use and portable (in terms of both operating systems and machine architecture). With the increasing growth in the sequence databases, it was also designed not to have limits on the number of sequences that can be included in an analysis or sequence length. Many codon-based analysis programs ignore the fact that there are alternatives to the universal genetic code. CodonW was designed to work with any variation of the genetic code. Seven alternatives to the universal genetic code have been built into CodonW; other codes can be added at compilation time, by editing a header file. Decisions regarding the translation of a codon, how synonymous each amino-acid is, the number of codons in a codon family, and which codons are synonyms are determined at run time. Although the COA output from CodonW is more easily interpreted when presented graphically, CodonW does not have in-built graphics. There are numerous programs specifically designed to work with numerical data (e.g. Cricket Graph (MAC or PC), StatsView (MAC), Excel (MAC, PC), Minitab (MAC, PC, VMS, UNIX), SAS (PC, UNIX, VMS), SPSS (PC, UNIX, VMS), Harvard Graphics (PC), gnuplot (X-windows)) and therefore graphical presentation of the data is left to external software. All the programs mentioned above accept ASCII input files, by default CodonW creates much of its output in a tabulated format designed to be easily read by eye (output from a COA is always machine-readable).

Codonw calculates statistical properties of genes such as codon adaptation index(CAI), frequency of optimal codons(Fop), codon bias index(CBI), Effective number of codons(NC), G+C content of the gene, G+C content 3rd position of synonymous codons(GC3s), Silent base compositions, Length silent sites(Lsil), Length of amino acids(Laa), and Aromacity score.

Significance

Synonymous codons are not used randomly and it has been previously shown that in many prokaryotes and some lower eukaryotes, that in addition to mutational biases, natural selection is also influencing this non-random codon usage. This selection is for a subset of "optimal" codons in those genes that are more highly expressed. The *insilico* analysis of codon usage of disease genes associated with human diseases is among the most critical tasks in medical research.

Codon mediated translational control may play an important role in the differentiation and regulation of tissue specific gene products and this can be linked to certain diseases in humans.

Evolutionary conservation rates, DNA coding lengths, gene functions of heritable disease genes and housekeeping genes may be compared. The human genome may contain thousands of essential as well as disease genes. The comparative account study will enable us to have clearer understanding of disease genes.

The abundance of enzymes in the disease gene set may reflect some historical bias towards metabolic disorders in the study of human inherited diseases.

Increasing knowledge of the genes associated with diseases will let us know and address important issues, including the relative contributions of disease genes in the core biological set and those encoding proteins specific to humans.

This study will also contribute in our understanding in why and how different genes evolve at different rates.

The outcome of this study can serve as a useful compendium for integrated information on disease genes with the potential for exploring druggabilty of disease genes.

References

Human disease genes, Gerado Jimenez-Sanchez, B. Childs and D. Valle; *Nature*,Vol 409, 15 FEB, 2001.

Tissue specific codon usage and expression of human genes, J.B. Plotkin, H. Robins & A.J. Levine; PNAS, 12588-12591, August 24, 2004.

Human disease genes: patterns and predictions, Nick G. C., Adam Eyre-Walker; GENE, 318(2003) 169-175.

OMIM, a knowledgebase of human genes and genetic disorders, A. Hamosh, A.F. Scott, J.S. Amerger, C.A. Bocchini & V.A. McKusick, *Nucleic Acid Research*. 2005,vol.33, Database issue.

Human housekeeping genes are compact, E. Eisenberg & E.Y. Levanon. *Trends in Genetics*, 19, 362-365(2003).

http://www.molbiol.ox.ac.uk/cu/DataRecording.html

23

Bioinformatics Tools for New Drug Designing and Development

K.K. Sardar, S.C. Parija and A.P.K. Mohapatra

In recent years, we have witnessed an explosion of biological information and the bioinformatics data arising from the rapid research and unprecedented progress in molecular biology. The novel genomic technologies have revolutionized the drug designing, discovery and development processes in government as well as private pharmaceutical industries through the use of these technologies. Various databases are doubling in size every year and fortunately we have the complete genome sequences of more than 100 organisms. At present, the pharmaceutical industries have clasped genomics as a source of drug targets. They also distinguish that the field of bioinformatics is crucial for validating these potential drug targets and for determining the ones that are most suitable for entering the drug development pipeline. Due to our increased understanding of molecular biology, there has been a change in the way that medicines are being developed. Once upon a time, new synthetic organic molecules were tested in animals or in whole organ preparations. Now this approach has been replaced with a molecular target in which *in-vitro* screening of compounds against purified, recombinant proteins or genetically modified cell lines is carried out with a high throughput as a consequence of better and ever improving knowledge of the molecular basis of disease.

At the moment, all marketed drugs do target only about 500 gene products. New opportunities for drug designing and discovery have been possible due mainly to the elucidation of the human genome which has an estimated 30,000 to 40,000 genes. The major advances in genomics and sequencing means that finding an attractive target is no longer a problem but finding the targets that are most likely to succeed has become the challenge. The focus of bioinformatics in the drug discovery process has therefore swung from target identification to target validation. A vast variety of factors need to be taken into account concerning a candidate target from a multitude of heterogeneous resources. The types of information that one needs to gather about potential targets include nucleotide and protein sequencing information, mapping information, pathway information, disease associations, variants, structural information, gene and protein expression data, etc. To gather the above information, different bioinformatics tools can be used. The accumulation of this information into databases about potential targets means that the pharmaceutical companies can save themselves much time, effort and expense exerting bench efforts. It is important that the potential target has a suitable structure for interacting with drug molecules. Structural genomics helps to prioritize the families in terms of their 3D structures. Every now and then we want to develop broad spectrum drugs that are effective against a wide range of pathogenic species while at other times we want to develop narrow spectrum drugs that are highly specific to a particular organism. Now, comparative genomics helps find protein families that are widely taxonomically dispersed and that are unique to a particular organism. In order to determine the task, we use DNA and protein chips to characterize the potential drug targets played in a particular disease mechanism. These chips can measure the amount of transcript or protein expressed by a cell at different times or in different states (healthy *versus* diseased). The field of bioinformatics has become a major part of the drug discovery pipeline and thus playing a key role for validating drug targets.

Use of Bioinformatics Tools for Drug Designing

Currently, the method of designing a new drug using bioinformatics tools have unlocked a new area of research. Though, computational techniques assist in searching drug target and in designing drug *in silico*, but it captures long time and money. In order to design a new drug one need to follow the following path.

- **Study Interesting Compounds**: One has to identify and study the lead compounds that have some activity against a disease. These may be only marginally useful and may have severe side effects. These compounds provide a starting point for refinement of the chemical structures.
- **Detect the Molecular Bases for Disease**: If it is known that a drug must bind to a particular receptor on a particular protein or nucleotide, then a drug can be made to bind at that site. This is often modeled computationally using any of several different techniques. Traditionally, the primary way of determining what compounds would be tested computationally was provided by the researchers' understanding of molecular interactions.
- **Identify Target Disease**: One requires to know about the disease and existing or traditional method of remedies. It is also important to distinguish

very similar sufferings and their rational treatments. Target identification alone is not sufficient in order to achieve a successful treatment of a disease. For a real drug to be developed, it must influence the target protein in such a way without interfering with normal metabolism processes of the organism. One way to achieve this is to block activity of the protein with a small molecule. Bioinformatics tools can be exploited to virtually screen the targets for compounds that bind and inhibit the protein. Other proteins that regulate the activity of the targets by binding and forming a complex may be another possibility for identifying the target.

- **Rational drug design techniques**: The complexity of biological systems makes it very difficult to determine the structures of large biomolecules. Ideally experimentally determined (x-ray or NMR) structure is desired, but biomolecules are very difficult to crystallize. The current techniques attempt to reproduce the researchers' understanding of how to choose likely compounds built into a software package that is capable of modeling a very large number of compounds in an automated way. Many different algorithms have been used for this type of testing, many of which were adapted from artificial intelligence applications.

- **Refinement of compounds**: Just the once you got a number of lead compounds, computational and laboratory techniques may be very successful in refining the molecular structures to give a greater drug activity and fewer side effects. This is done both in the laboratory and computationally by examining the molecular structures to determine which aspects are responsible for both the drug activity and the side effects. This can be done using 'quantitative structure activity relationships (QSAR)' that consists of computing every possible number that can describe a molecule then doing an enormous curve fit to find out which aspects of the molecule correlate well with the drug activity or side effect severity

- **Solubility of Molecule**: One needs to confirm whether the target molecule is lipid soluble or water soluble or readily soluble in fatty tissue. The ability to get a drug to the correct part of the body is an important factor in its potency. Ideally there is a continual exchange of information between the researchers doing QSAR studies, synthesis and testing. These techniques are frequently used and often very successful since they do not rely on knowing the biological basis of the disease which can be very difficult to determine.

- **Drug Testing on patients**: Once a drug has been demonstrated to be effective by an initial assay technique, much more testing must be done before it can be given to human patients. Animal testing is the primary type of testing at this stage. Eventually, the compounds, which are deemed suitable at this stage, are sent on to four phases of clinical trials.

Computer-Aided Drug Design

It is always key to appreciate the functioning of drugs before getting into the molecular aspects, because drug can interfere with or enhance our own biological function and can block interaction of bacteria, viruses and parasites in our system.

Pharmacogenomics: It is central to know that drug designing includes pharmacogenomics (gene-based design of diagnostics and therapeutics) and drug molecule designing. Now, pharmacogenomics is the application of genomic approaches and technologies to the identification of drug targets. For example, in potential cancer therapy targets, we have to travel entire genomes for potential receptors by bioinformatics means, or investigate patterns of gene expression in both pathogens and hosts during infection, or examine the characteristic expression patterns found in tumours or patients samples for diagnostic purposes.

Pharmacogenetics

Pharmacogenetics being a subset of pharmacogenomics uses genomic/ bioinformatics methods to identify genomic correlates, for example SNPs (Single Nucleotide Polymorphisms), characteristic of particular patient response profiles and use of those markers to inform the administration and development of therapies. Strikingly such approaches have been used to "resurrect" drugs thought previously to be ineffective, but subsequently found to work with in subset of patients or in optimizing the doses of chemotherapy for particular patients. Individual differences exist to drug therapy; some positively, others with little obvious change in their conditions and yet others with side effects or allergic reactions. Much of this variation is known to have a genetic basis of life.

Cheminformatics

Penicillin, being a chance-finding drug, was discovered by Sir Alexander Flemming in 1928. The discovery of Penicillin is thus a very successful story and the way we discover and develop drugs even now has similarities to this discovery. Until recently, drug design always seems to be a labour-intensive, trial-and-error process. Information technology can now be effectively used by chemists and biochemists to plan intelligently and to automate processes related to the chemical synthesis of possible therapeutic compounds. The rewards for bringing a drug to market more rapidly are huge, so naturally this is what a lot of cheminformatics works viz., Synthesis Planning • Reaction and Structure Retrieval • 3-D Structure Retrieval • Modelling • Computational Chemistry.

Drug Development

More specific medicines that act on the cause, not merely the symptoms, of the disease can be developed with an improved understanding of disease mechanisms and using computational tools to identify and validate new drug targets. These highly specific drugs promise to have fewer side effects than many of today's medicines. Scientists have been examining the genome of *Enterococcus faecalis*-a leading cause of bacterial infection among hospital patients. They have discovered a virulence region made up of a number of antibiotic-resistant genes that may contribute to the bacterium's transformation from a harmless gut bacteria to a menacing invader. The discovery of the region, known as a *pathogenicity island*, could provide useful markers for detecting pathogenic strains and help to establish controls to prevent the spread of infection.

Veterinary Science Till now, sequencing projects of many farm animals including cows, sheep and pigs are now well under way in the hope that a better understanding of the biology of these organisms will have huge impacts for improving the production, reproduction and health of livestock and ultimately have benefits and welfare for human health and nutrition.

Bioinformatics-Computer-aided Drug Development

Every now and then, we have a little or more of the recipe of life stretching before us as endless line of adenine, guanine, cytosine and thymine-forming general sequences common to most living organisms. A vital thrust of 21st century biology will be the animation of these static parts. The field of drug discovery and design has been enriched by the genomic technologies viz. Novel sequencing technologies, DNA chip technology or micro array technology, Proteomic technology, and metabolic profiling technologies. These technologies have created profound impact on pharmacology, medicine and diagnostics. Now-a-days, computers are an integral part of genomics-based drug discovery, helping researchers find drug targets by comparing databases of genomic information with annotations about functional information, by analyzing the data that comes in from various wet lab experiments and by simply keeping track of the huge amounts of biological data being uncovered in life sciences research. This is the role of bioinformatics, a field that has exploded in importance over the last few years as companies have begun to realize they are drowning in raw data.

While academic labs try to create accurate simulations of red blood cells and simple bacteria, the private companies are taking on bolder projects—simulating human organs and even human diseases in their entirety. Theoretically, researchers could now test virtual drug compounds against virtual protein targets, study the virtual pharmacokinetics of their optimized virtual lead in what amounts to virtual animals, study its effects on virtual organs, design a virtual clinical trial to test assumptions and variances, and even answer some regulatory questions through simulation. Somewhere in that process, a chemist has to actually mix up a compound and conduct some experiment but buckets of silicon are being added to the discovery and development process every day, with the hope that the wet lab will one day become as dry as a sand box. One way of animating our growing store of static information is through computer simulation. It is an area that is beginning to emerge slowly in the life sciences, with only a handful of academic and commercial players active in the area. But for a fledging discipline, there is a great variety in the scope of work being undertaken.

24

In Silico Drug Designing

B.P. Dash

Drug designing is the approach of finding drugs by design based on their biological targets. The substance used for medical purpose, either alone or in mixture that changes the state or function of cells, organs or organism as a whole is known as a drug. So typically a drug target is a key molecule involved in particular metabolic or signaling pathway that is specific to disease condition or pathology, or to the infectivity or survival of a microbial pathogen. Drugs may be designed that bind to the active region and inhibit the key molecule that causes disease. However these drugs would also have to be designed in such a way that they should not affect any other important molecules similar to the key molecule. These can also be used to enhance the normal pathway of metabolism by promoting specific molecules that may have been suppressed in disease state.

Can Drug be Designed?

The need for ongoing development of new drugs needs no emphasis in light of the current global situation of health and disease. Traditionally, the process of drug development has revolved around a screening approach, as nobody knows which compound or approach could serve as a drug or therapy. Such almost blind screening approach is very time-consuming and laborious. The shortcoming of traditional drug discovery; as well as the allure of a more deterministic approach to combating disease has led to the concept of "Rational drug design". Nobody could design a drug before knowing more about the disease or infectious process than past. For "rational" design, the first necessary step is the identification of a molecular target critical to a disease process or an infectious pathogen. Then the important prerequisite of "drug design" is the determination of the molecular structure of target, which makes sense of the word "rational". The validity of "rational" or "structure-based" drug discovery rests largely on a high-resolution target structure of sufficient molecule detail to allow selectivity in the screening of compounds. When the above information is available, the "rational" design of drug will be more possible.

What is Drug Design?

Drug discovery and development is an intense, lengthy and an interdisciplinary endeavor. Drug discovery is mostly portrayed as a linear, consecutive process that starts with target and lead discovery, followed by lead optimization and pre-clinical *in vitro* and *in vivo* studies to determine if such compounds satisfy a number of pre-set criteria for initiating clinical development. For the pharmaceutical industry, the number of years to bring a drug from discovery to market is approximately 12-14 years and costing up to \$1.2 - \$1.4 billion dollars. Traditionally, drugs were discovered by synthesizing compounds in a time-consuming multi-step processes against a battery of *in vivo* biological screens and further investigating the promising candidates for their pharmacokinetic properties, metabolism and potential toxicity. Such a development process has resulted in high attrition rates with failures attributed to poor pharmacokinetics (39%), lack of efficacy (30%), animal toxicity (11%), adverse effects in humans (10%) and various commercial and miscellaneous factors. Today, the process of drug discovery has been revolutionized with the advent of genomics, proteomics, bioinformatics and efficient technologies like, combinatorial chemistry, high throughput screening (HTS), virtual screening, *de novo* design, *in vitro*, *in silico* ADMET screening and structure-based drug design.

What is *In Silico* Drug Design?

In silico methods can help in identifying drug targets via bioinformatics tools. The use of computers and computational methods permeates all aspects of drug discovery today and forms the core of structure-based drug design. High-performance computing, data management software and internet are facilitating the access of huge amount of data generated and transforming the massive complex biological data into workable knowledge in modern day drug discovery process. The use of complementary experimental and informatics techniques increases the chance of success in many stages of the discovery process, from the identification of novel targets and elucidation of their functions to the discovery and development of lead compounds with desired properties. Computational tools offer the advantage of delivering new drug candidates more quickly and at a lower cost. Major roles of computation in drug discovery are; (1) Virtual screening & de novo design, (2) *in silico* ADME/T prediction and (3) Advanced methods for determining protein-ligand binding.

Significance of *In Silico* Drug Design

As structures of more and more protein targets become available through crystallography, NMR and bioinformatics methods, there is an increasing demand for computational tools that can identify and analyze active sites and suggest potential drug molecules that can bind to these sites specifically. Also to combat life-threatening human diseases such as AIDS, Tuberculosis, Malaria etc. or agricultural diseases or pathogens, a global push is essential.

Structure Based Drug Design

The crystal structure of a ligand bound to a protein provides a detailed insight into the interactions made between the protein and the ligand. Structure designed can be used to identify where the ligand can be changed to modulate the physicochemical and ADME properties of the compound, by showing which parts of the compound are important to affinity and which parts can be altered without affecting the binding. The equilibrium between target and ligand is governed by the free energy of the complex compared to the free energy of the individual target and ligand. This includes not only the interaction between target and ligand but also the solvation and entropy of the three different species and the energy of the conformation of the free species. As structures of more and more protein targets become available through crystallography, NMR and bioinformatics methods has bring a major drive in the computational methods to use the structure of the protein target as a route to discover novel lead compounds. The methods include de novo design, virtual screening and fragment based discovery.

Virtual screening and de novo design play an important role within the pharmaceutical industry in lead discovery process. Virtual screening refers to computational screening of large libraries of chemicals for compounds that complement targets of known structure which could be tested experimentally. Since, the virtual screening takes place in the three-dimensional active site of the target; it is also called as structure-based virtual screening.

De novo design attempts to use the unliganded structure of the protein to generate novel chemical structure that can bind. There are varying algorithms, most of which depend on identifying initial putative sites of interaction that are grown into complete ligands.

Fragment based discovery is based on the premise that most ligands that bind strongly to a protein active site can be considered as a number of smaller fragments or functionalities. Fragments are identified by screening a relatively small library of molecule (400-20,000) by X-ray crystallography, NMR spectroscopy. These structures of the fragment binding to the protein can be used to design new ligands by adding functionality to the fragments or by incorporating features of the fragment onto existing ligands.

Conclusion

In silico drug designing is now a powerful strategy to design a drug of clinical/ biological or agricultural importance. Development of potential drugs to check or cure some diseases of plants or crops is of ought most urgency. A good number of either of crops or human diseases is still rampant which needs a permanent solution. Due to the bottleneck in traditional drug designing a success in this endeavor is not in sight. But due to the rapid development of computer based bioinformatics knowledge on biological data accumulation, interpretation and manipulation for in silico drug designing will definitely solve to achieve the desired target in near future.

25

Avian flu and Vaccine Design

R. Nayak

Influenza a Pandemics

Pandemics	Year of pandemic	Influenza type
Italy:a disease identified as "fowl plague"-avian influenza and new castle disease	1878	H3N8
Northern Europe and United States	1890	H2N2
Spanish flu	1918	H1N1
Differentiation of the two diseases	1927	H3N8
First strains isolated	1933	H1N1
Variation detected	1947	H1N1
Asian flu pandemic	1957	H2N2
Hong Kong flu pandemic	1968	H3N2
Swine flu, non-epidemic	1976	H1N1
Russian pandemic	1977	H1N1+H3N2
9 South-East Asian countries	2004-2006	H5N1

H5N1 – causes avian influenza in humans, chickens, goose, ducks and many migratory birds.

Bird to Bird Transmission- mostly seen

Bird to Human Transmission- rarely seen

Human to Human Transmission- normally not seen

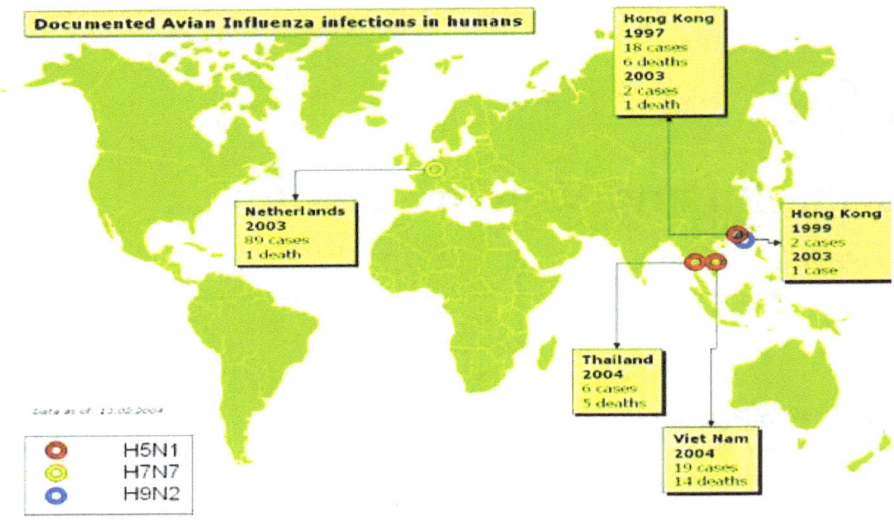

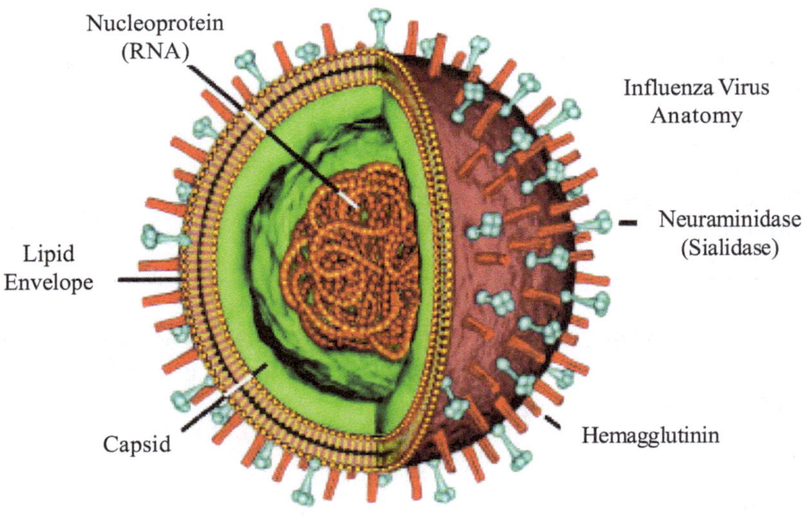

Structure of Influenza A Virus

Surface Glycoproteins

- Hemagglutinin- 16 Subtypes
- Neuraminidase- 9 Subtypes

Protein names	Gene length	No. of peptides generated
Hemagglutinin	568	560
Neuraminidase	455	457
Matrix protein 1	252	254
Matrix protein 2	93	85
Nucleocapsid protein	498	490
Polymerase	716	708
Polymerase basic protein 1	757	749
Polymerase basic protein 2	759	751
Nonstructural protein 1	225	217
Nonstructural protein 2	111	103

Influenza virus

Family- Orthomyxoviridae

Negative sense, single stranded, RNA virus

8 Genome segments

11 Proteins [including HA, NA, NS1, NS2, M1, M2, RNP-PA, PB1, PB2 & NP]

Receptors

- *-2,3 Glycosialic acid- Birds*
- *-2,6 Glycosialic acid- Humans*

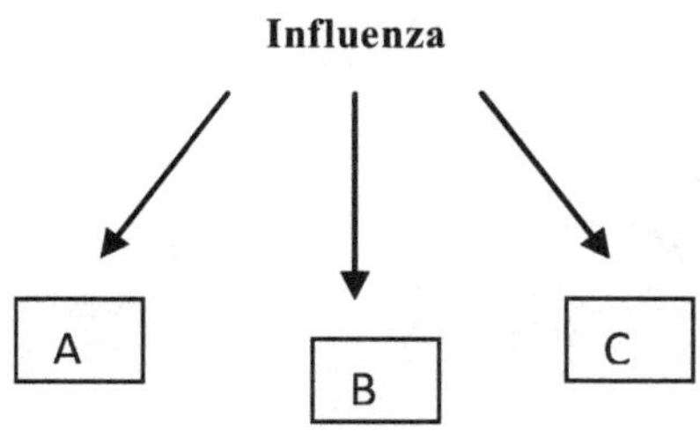

Influenza A viruses- humans, ducks, chickens, pigs, whales, horses and seals.

Influenza B viruses- widely only among humans.

Influenza C viruses- humans rare.

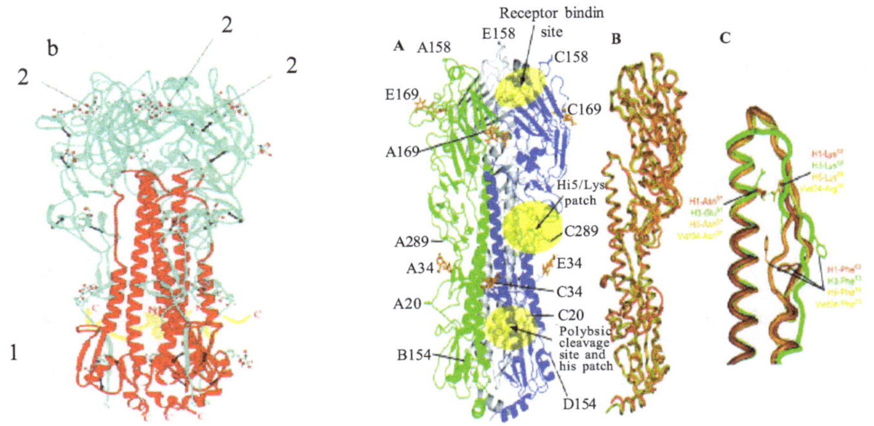

Cleaved BHA (3). Receptor-binding sites marked with *circled 2* (4)

Problems in Vaccine Development

Antigenic drift

Antigenic Shift

Variation of Influenza Viruses

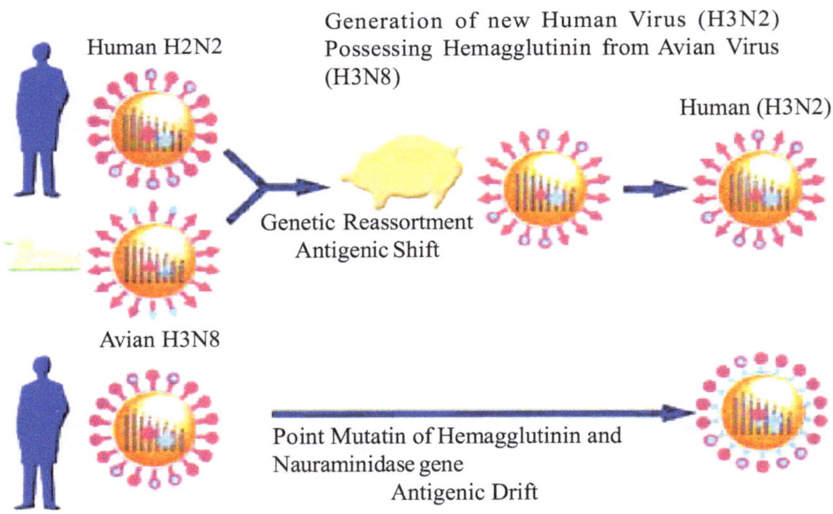

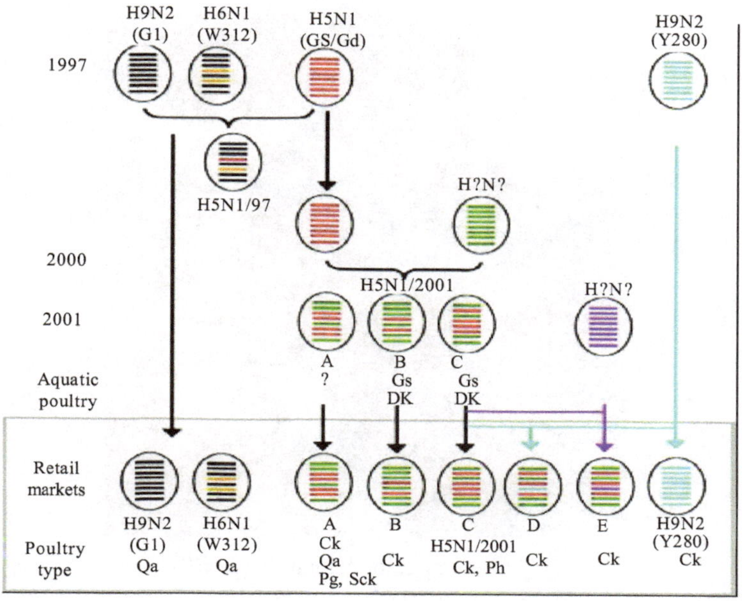

1. Mutation changes the antigenic site.
2. Antibody against original influenza.
3. Influenza neutralize mutated virus.
4. Influenza is not prevented in spite of vaccination.

So, vaccine must be prepared against the organism of the infection cycle.

Difficulties in Developing Seasonal Vaccines

- Logistics
- Cost
- Time
- Materials

Who will be vaccinated???

Human! or Birds!

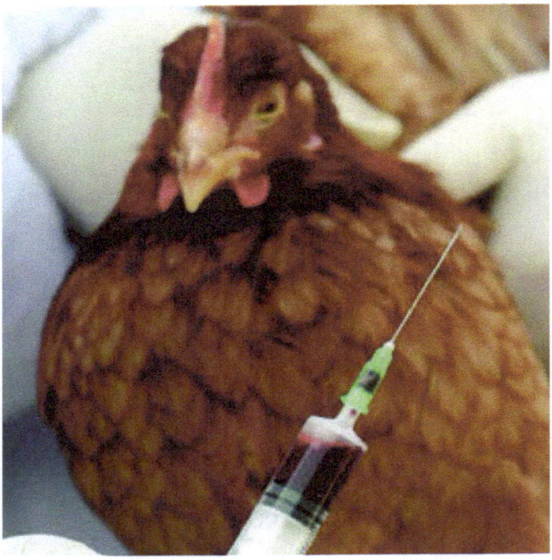

Immunization of personnel handling birds

Strategy for Vaccine Design

1. Analyse the protein of the pathogen for putative B- and T-Cell Epitope by computational methods.
2. Find haplotype specificity of peptides.
3. Find the selfness with the host protein by blasting the putative epitopes with Human Proteome.
4. Experimentally verify Hla binding In *In Vitro* assays.
5. Experimentally verify the immune response generated by peptides in MHC compatible animal model.
6. Select appropriate peptides for construction of chimeric vaccine.
7. Test the vaccine in animal model.
8. Human trials

HLA polymorphism in humans (mar 08)

HLA Class I										
Gene	A	B	C	E	F	G	H	J	K	L
Alleles	451	782	238	9	20	23	0	0	0	0
Proteins	358	672	188	3	4	6	0	0	0	0
Nulls	36	25	6	0	0	1	0	0	0	0

HLA Class II										
Gene	DRA	DRB	DQA1	DQB1	DPA1	DPB1	DMA	DMB	DOA	DOB
Alleles	3	522	34	71	23	124	4	7	12	9
Proteins	2	430	25	55	14	111	4	7	3	4
Nulls	0	7	1	1	0	2	0	0	1	0

Steps involved in vaccine development:

Work carried out

Prediction of MHC class I bindin epitopes
↓
Prediction of MHC Class II binding epitopes
↓ 236 Peptides
Determination of promiscnous epitope binding peptides
↓ 190 peptides
B-cell epitope mapping (Kyte Doolittle plot)
↓ 134 peptides selected
Proteosome cleavage-Fragpredict
↓ 51 peptides selected
Conserved Regions
↓ 47 Peptides
Desired Peptides
↓ 37 peptides
Energy minimization-CNS
↓

Remains to be done

Experimentally verify HLA Binding in *in vitro* Assays
↓
Experimentally verify the immune Response generated by peptides in MHC Compatible Animal Model
↓
Select Appropriate Peptides for Construction of Chimeric Vaccine
↓
Test the vaccine in animal Model (Ethical Clearancea)
↓
Human testing (Ethical clearance)

A computational approach to define T-cell epitopes of H5N1 Avian influenza virus with vaccine potential

MHC class-I peptide prediction tools

Analytical Tools	Basis of Analysis	Optimal Value	No.of Alleles (Nonamers)
Bimas	Experimental Analysis	T1/2 ≥ 100	33
Syfpeithi	Experimental Analysis	SCORE ≥15	14
Rankpep	Protein Specific Scoring Matrices(PSSMS)	Binding Threshold 2%	72

MHC Class II Peptide Predictions Tools

Analytical Tools	Basis of Analysis	Optimal Value	No.of Alleles (Nonamers)
PROPRED	Amino acid/position co-efficient table	Binding threshold 3%	51
RANKPEP	Protein specific scoring matrices (PSSMS)	Binding threshold ~5%	49

Promiscuous Epitope

Promiscuous epitope is defined as the peptides having the capacity to bind to more than one allele. From the analysis done by class I MHC binding predictions, peptides binding to more than two alleles were chosen for the promiscuous epitope mapping. The promiscuousity was checked at various cut off scores i.e., at 1, 10, 50 and 100. This was done to check the stringency of the peptides binding to different alleles at various t½ dissociation rates.

Human-Self Index

The Hu-SIN of a pathogen peptide is defined as the number of continuous amino acids matching with that of the human peptides divided by the total number of nonameric peptides. The Hu-SIN is calculated according to the formula:

$$\text{Hu-SIN}(\%) = \frac{\text{No. of continuous amino acids in pathogen showing homology with human peptide}}{\text{No. of amino acids in the peptide}} \times 100$$

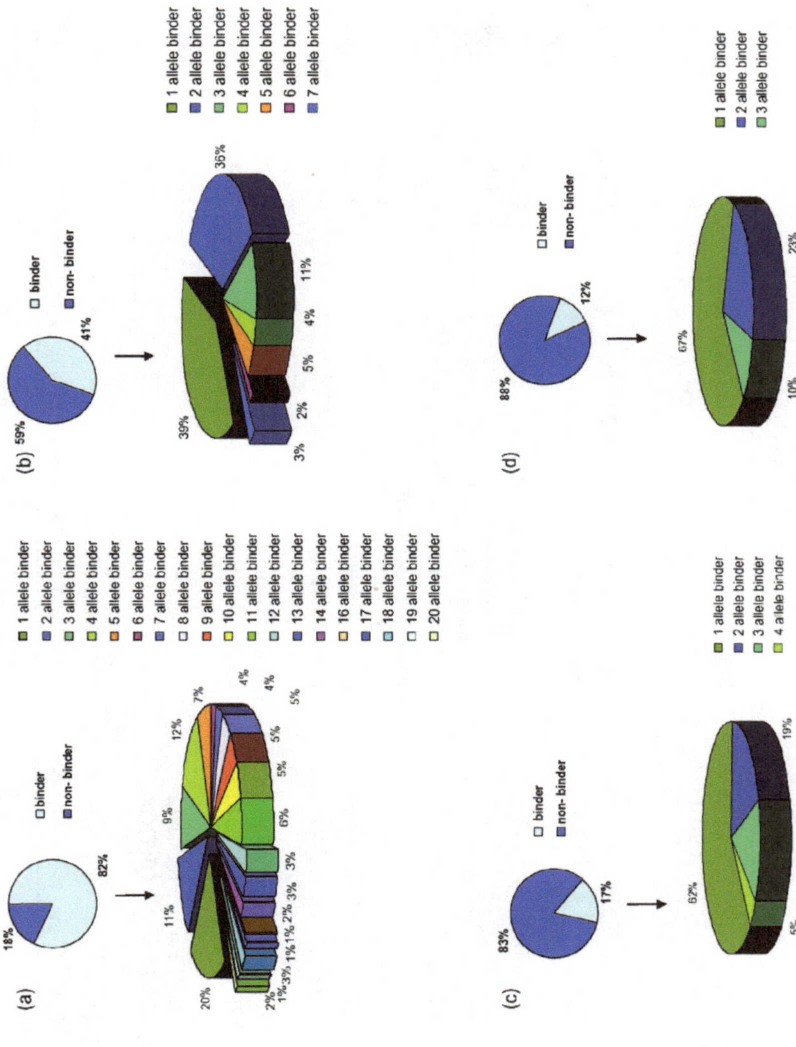
(a)
18%
82%
binder
non- binder
1 allele binder
2 allele binder
3 allele binder
4 allele binder
5 allele binder
6 allele binder
7 allele binder
8 allele binder
9 allele binder
10 allele binder
11 allele binder
12 allele binder
13 allele binder
14 allele binder
16 allele binder
17 allele binder
18 allele binder
19 allele binder
20 allele binder
(b)
59%
41%
binder
non- binder
39%
35%
1 allele binder
2 allele binder
3 allele binder
4 allele binder
5 allele binder
6 allele binder
7 allele binder
(c)
83%
17%
binder
non- binder
62%
19%
14%
5%
1 allele binder
2 allele binder
3 allele binder
4 allele binder
(d)
88%
12%
binder
non- binder
67%
23%
10%
1 allele binder
2 allele binder
3 allele binder

Table 2: Representative HLA binding peptides of different proteins of Avian influenza sharing homology with human proteins.

Protein	Viral Peptide	Human Peptide	Human Protein Name	Hu-SIN	T(½) (minutes)	Binding HLA
HA	VLMENERTL	MENERT	high density lipoprotein binding (HDLBP)	67.00	110	A_0201
	MPFHNIHPL	FHNIHP	acyl-CoA synthetase	67.00	286, 666, 160	B_5101, _5102, Cw_0401
NA	QEFKMNPNK	FKMNPN	cadherin protein, dachsous 2 isoform 1	67.00	150	B_2705
M I	RRRFVQNAL	RRRFV	neurobeachin	56.00	180, 6000	B_2702, B_2705
	ARQMVQAMR	QMVQAM	UDP-galactose-4-epimerase	67.00	1000	B_2705
NS I	DRLRRDQKS	RLRRDQK	Transducin beta-like 3	78.00	150	B14
	LEEESDKAL	EEESDKA	Hypothetical protein	78.00	160	B60
NS II	GMITQFESL	MITQFE	similar to kinesin family member 18A isoform 6	67.00	100	Cw_0301
PA	YYLEKANKI	YLEKANK	Dystrophin	78.00	119	A24
	HEKGINPNY	EKGINPN	similar to defens in	78.00	180	B_4403
Protease	YQVSLNSGY	YQVSLNSGY	trypsin-like protein	100.00	160100	B62, B_2705
	VVSAAHCYK	VVSAAHCYK	HGF activator, prepro protein	100.00	120	A68.1
NC	LQNSQVFSL	QNSQVFS	similar to kinesin-like motor protein C20 or f23	78.00	200	B_2705
PB I	MYQKCCNLF	KCCNLF	zinc finger protein	67.00	180100	A24,
Cw _0401	MVTQRTIGK	QRTIGK	fragile X mental retardation protein	67.00	120	A68.1
PB II	ITKEKKEEL	TKEKK	DNA topoisomerase II	56.00	120	B8
	VVLVMKRKR	LVMKRKR	zinc finger protein, protein-tyrosine phosphatase	78.00	400	A68.1

Conserved Regions

Various strains of Influenza type A like H5N1, H3N2, H2N2, H1N1, H3N8 and H7N7 were aligned using clustalw multiple alignment algorithm. This will allow us to focus on the conserved regions in different protein sequences of the avian virus proteome, where there is less probability of mutations/antigenic drift.

H5N1 Hem agglutinin protein contains predicted immunogenic peptides with potential to induce autoimmune reaction: implication for vaccine development.

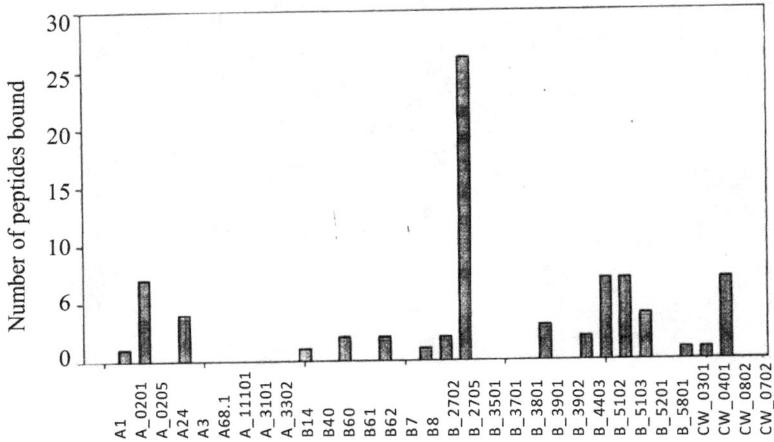

MEKIVLLFAIVSLVKSDQICIGYHANNSTEQVDTIMEKNV 40
TVTHAQDILEKTHNGKLCDLDGVKPLILRDCSVAGWLLGN 80
PMCDEFINVPEWSYIVEKANPVNDLCYPGDFN*DYEELKHL* 120
LSR*INHFEK*IQIIPKSSWSSHEASLGVSSACPYQGKSSFF 160
RNVVWLIKKNSTYPTIKRSYNNTNQEDLLVLWGIHHPNDA 200
AEQTKL*YQNPTTY*ISVGTSTLNQRLVPRIATRSKVNGQSG 240
RMEFFWTILKPNDAINFESNGNF*IAPEYA*YKIVKKGDSTI 280
MKSEL*EYGNC*NTKCQTPMGAINSSMP*FHNIHP*LTIGECPK 320
YVKSNR*LVLATGL*R*NSPQRERRRKK*RGLFGAIAGFIEGGW 360
QGMVDGWYGYHHSNEQGSGYAADKESTQKAIDGVTNKVNS 400
IIDKMNTQFEAVG*REFNNLERRIENLNK*KMEDGFLDVWTY 440
NAELLVL*MENERT*LDFHDSNVKNLYDK*VRLQLRD*NAKELG 480
*NGCFEF*YHKCDNEC*MESVRN*GTYDYPQ*YSEEARLKREEIS* 520
*GVKLESIG*IYQILSIYSTVASSLALAIMVAGLSLWMCSNG 560
SLQCRICI 568

The conserved epitopes of internal proteins of influenza virus are defined to develop vaccine which would overcome the problems associated with antigenic drift and antigenic shift.

26

Earthquake Resistant Design of Agricultural Structures By Use of Software

L. Gouda

This statement signifies that the casualty of people is not due to the earthquake but due to the damage of buildings by earthquake forces. In developed countries, the casualty and loss of property is very minimal in comparison to the death toll and losses in India in the event of earthquakes of same magnitude. It has been experienced in the past that on the event of earthquakes of Richter magnitude scale of 6 or more, the death toll in India is in terms of thousands where as in Western Countries, it is multiple of tens. The less casualty is due to less damage of buildings and other infrastructures which in turn is due to the design and construction of Buildings and other structures as per the earthquake resistant theory. This paper discusses the issues related to the earthquake hazards in India and the concept of earthquake resistant structures.

Earthquake hazard zoning of India

The Indian subcontinent had a long history of devastating earthquakes. The major reason for the high frequency and intensity of the earthquakes is that India continues to drive into Asia at a rate of approximately 47 mm/year. Geographical statistics of India show that almost 54% of the land is vulnerable to earthquakes. The latest version of seismic zoning map of India(Fig.1) given in the earthquake resistant design code of India [IS 1893 (Part 1) 2002] assigns four levels of seismicity for entire India in terms of different zone factors. In other words, the earthquake zoning map of India divides India into 4 seismic zones (Zone 2, 3, 4 and 5) unlike its previous version which consisted of five or six zones for the country. According to present zoning map, Zone-5 expects highest level of seismicity whereas Zone-2 is associated with lowest level of seismicity.

The MSK (Medvedev-Sponheuer-Karnik) intensity broadly associated with the various seismic zones is VI (or less), VII, VIII and IX (and above) for Zones-2, 3, 4 and 5, respectively, corresponding to Maximum Considered Earthquake (MCE). The IS code follows a dual design philosophy: (a) under low probability or extreme earthquake events (MCE) the structure damage should not result in total collapse, and (b) under more frequently occurring earthquake events, the structure should suffer only minor or moderate structural damage. The specifications given in the design code (IS 1893: 2002) are not based on detailed assessment of maximum ground acceleration in each zone using deterministic or probabilistic approach. Instead, each zone factor represents the effective period peak ground accelerations that may be generated during maximum considered earthquake ground motion in respective zone. Each zone indicates effects of an earthquake at a particular place based on the observations of the affected areas and can also be described using a descriptive scale like Modified Mercalli intensity scale or the Medvedev-Sponheuer-Karnik scale.

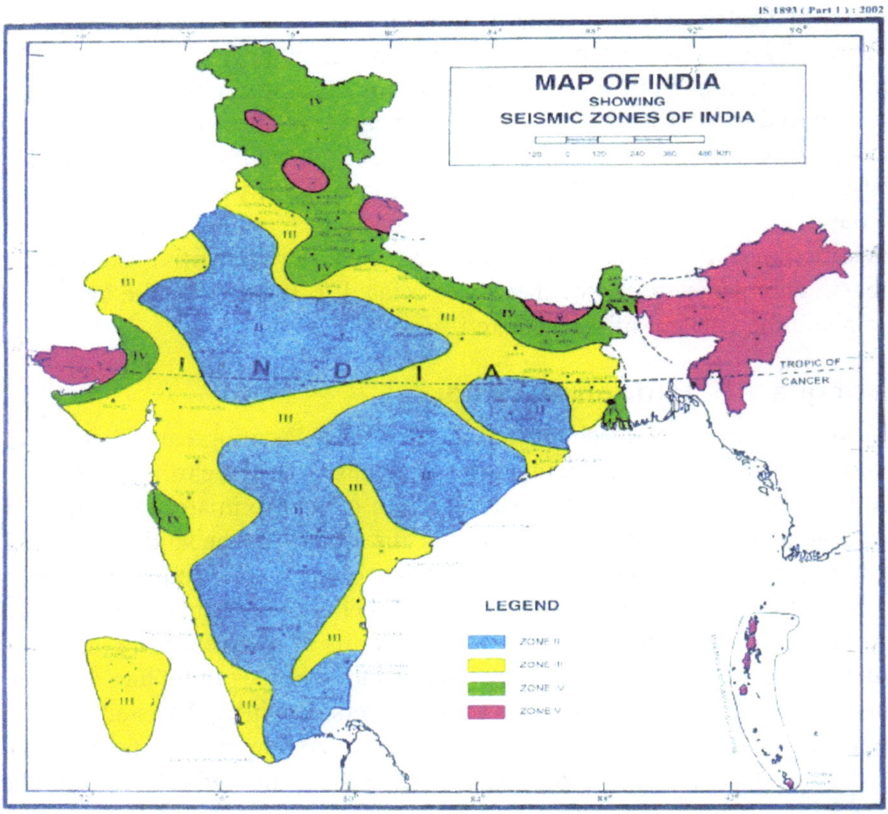

NOTE : Towns falling at the boundary of zones demarcation line between two zones shall be considered in High Zone.

Fig. 1: Seismic zoning map of India

Zone 5

Zone 5 is the worst prone zone in the country. It covers the area with the highest risk zone that suffers earthquakes of intensity MSK IX or greater and **Richter magnitude scale greater than 8.** The IS code assigns zone factor of 0.36 for Zone-5. Structural designers use this factor for earthquake resistant design of structures in zone-5. The zone factor of 0.36 is indicative of effective (zero period) peak horizontal ground accelerations of 0.36 g (36 % of gravity) that may be generated during MCE level earthquake in this zone. It is referred to as the **Very High Damage Risk Zone**. The state of Kashmir, Punjab, the western and central Himalayas, the North-East Indian region and the Rann of Kutch fall in this zone. Generally, the areas having trap or basaltic rock are prone to earthquakes.

Zone 4

This zone is called the High Damage Risk Zone and covers areas liable to MSK VIII. And Richter magnitude scale 7 to 7.9.The IS code assigns zone factor of 0.24 for Zone-4. The Indo-Gangetic basin and the capital of the country(Delhi, Jammu) fall in Zone 4.

Zone 3

The Andaman and Nicobar Islands, parts of Kashmir, Western Himalayas fall under this zone. The Killari (Latur) area, Narmada Tectonic Domain, Eastern coast area (Bhubaneswar & Cuttack), Mahanadi Graben and Godawari Graben area (Vijaywada, Nellore & Chennai) also fall in this zone. This zone is classified as Moderate Damage Risk Zone which is liable to MSK VII. **Richter magnitude scale 6 to 6.9.**The IS code assigns zone factor of 0.16 for Zone-3.

Zone 2

This region is liable to MSK VI or less and **Richter magnitude scale less than 6** and is classified as the Low Damage Risk Zone. The IS code assigns zone factor of 0.10 (maximum horizontal acceleration that can be experienced by a structure in this zone is 10 % of gravitational acceleration) for Zone-2.

Earthquakes in India

The list of damaging earthquakes in India is given in the table-1.The table shows that there were serious earthquakes of high magnitude causing severe damage to the life and property. In the recent past in the year 2001, there was an earthquake of magnitude 6.9 on Richter scale whose epicenter was located near Bhachau, widely known as Bhuj earthquake. This earthquake has given us some lessons. It was found that government buildings constructed by Central Public Works Department CPWD) performed relatively better than other private buildings on account of construction of these buildings in compliance to the provisions of the codes IS 1893(Part-1):2002 and IS 13920:1993.

Table 1: List of damaging earthquakes in India

S.No.	Name of earthquake	Location	Year	Magnitude/ Intensity	Death
Himayalayan Region					
1.	Kashmir	Srinagar, J & K	1885	-	3000
2.	Shillong	Shillong	1897	8.7	1600
3.	Kangra	Kangra, H.P.	1905	8.5	20,000
4.	Bihar-Nepal	Bihar- Nepal border region	1934	8.3	10,000
5.	Assam	Assam	1950	8.5	1,526
6.	Bihar-Nepal	Bihar-Nepal border region	1988	6.5	1,000
7.	Indo-Burma	India-Burma border	1988	7.3	-
8.	Uttarkashi	Uttarkashi, Uttaranchal	1991	7.0	768
9.	Chamoli	Chamoli, Uttaranchal	1999	6.8	103
Andaman Nicobar Region					
1.	Andaman-Nicobar	Andaman-Nicobar Trench	1941	8.1	-
Kutch Region					
1.	Samaji	Samaji, Delta of Indus	1668	X	-
2.	Kutch	Kutch, Gujarat	1819	8.0	2,000
3.	Anjar	Anjar,Gujarat	1956	6.1	115
4.	Bhuj	Bachau,Gujarat	2001	6.9	20,000
Region of Peninsular India					
1.	Bombay-Surat	Bombay-Surat	1856	VII	-
2.	Son Valley	Son Valley	1927	6.5	-
3.	Satpura	Satpura	1938	6.3	-
4.	Balaghat	Balaghat, M.P.	1957	5.5	-
5.	Koyna	Koyna	1967	6.0	177
6.	Ongole	Ongole, Bhadrachalam	1967	5.4	-
7.	Broach	Broach	1970	5.4	26
8.	Latur	Latur, Maharashtra	1993	6.2	10,000
9.	Jabalpur	Jabalpur, M.P.	1997	6.0	54

Earthquake Resistant Structures

IS 1893(part-1) : 2002 describes the criteria for earthquake resistant design of structures. IS 13920-1993 describes about the ductile detailings of reinforced concrete structures subjected to seismic forces. It has been experienced that structures designed as per the provisions in the above said codes have performed better in the event of earthquakes. IS 4326, IS 13827, IS 13828, IS 13920 and IS 13935 describe the provisions about the construction. Government of India, Home ministry has taken-up initiatives in this regard. National training programs for the capacity building of the Engineers and Architects for design and construction of earthquake resistant structures are being organized in different parts of the country. The Farm Houses, ware houses, Cold storages and Go-downs are very important and costly structures for the agricultural developments. We are facing shortage of these storage facilities. Many parts of our country are vulnerable to earthquake. So these structures need to be designed and constructed as earthquake resistant structures for their sustainable development.

Softwares for design of Earthquake Resistant Structures

Earthquake imposes a dynamic load on the structure whose analysis involves quite a good number of second order equations. So a number of softwares have been developed by different companies worldwide. Some of such softwares are STAAD-Pro, SAP, STRUDS, BASE, SAFE, RISA-3D, STRAN, NESTRUM, NISA CIVIL which are widely accepted and used by Structural Engineers for design of earthquake resistant structures within a very less time period.

Conclusion

All the structures should be designed and constructed as earthquake resistant structures. The human civilization is facing a great challenge to reduce or mitigate the adverse effects of natural disasters, particularly due to earthquakes. Technology has been developed by conducting extensive and expensive research works. Softwares are readily available which are rendering a great help in designing any type of earthquake resistant structure. It is pertinent to mention that the post disaster effects can be highly minimized as the damages to the structures and the losses to the life and property can be greatly reduced by construction of earthquake resistant structures.

Index

D

E

F

G

H

I

L

M

N

O

P

U

V

W

X